D0867215

RESERVOIR LIMNOLOGY

RESERVOIR LIMNOLOGY: ECOLOGICAL PERSPECTIVES

Edited by

Kent W. Thornton
FTN Associates, Ltd.
Little Rock, Arkansas

Bruce L. Kimmel
Environmental Sciences Division
Oak Ridge National Laboratory
Oak Ridge, Tennessee

Forrest E. Payne
FTN Associates, Ltd.
Little Rock, Arkansas

A Wiley-Interscience Publication
John Wiley & Sons, Inc.
New York / Chichester / Brisbane / Toronto / Singapore

***Library of Congress Cataloging in Publication Data*:**

Reservoir limnology/edited by Kent W. Thornton, Bruce L. Kimmel,
Forrest E. Payne.
p. cm.
Includes bibliographical references.
ISBN 0-471-88501-0
1. Reservoir ecology. 2. Limnology. I. Thornton, Kent W.
II. Kimmel, Bruce L. III. Payne, Forrest E.
QH541.5.R4R47 1990
574.5′26322—dc20 89-24860
CIP

Printed in the United States of America

10 9 8 7 6 5 4 3 2

CONTRIBUTORS

Thomas M. Cole
Environmental Laboratory, Department of the Army, Waterways Experiment Station, Corps of Engineers, Vicksburg, Mississippi

Dennis E. Ford
Principal Hydrologist, FTN Associates, Ltd., Little Rock, Arkansas

Herbert H. Hannan
Aquatic Station, Biology Department, Southwest Texas State University, San Marcos, Texas

Robert H. Kennedy
Environmental Laboratory, Department of the Army, Waterways Experiment Station, Corps of Engineers, Vicksburg, Mississippi

Bruce L. Kimmel
Environmental Sciences Division, Oak Ridge National Laboratory, Oak Ridge, Tennessee

Owen T. Lind
Department of Biology and Center for Environmental Studies, Baylor University, Waco, Texas

G. Richard Marzolf
Hancock Biological Station, Murray State University, Murray, Kentucky

W. John O'Brien
Department of Ecology and Systematics, University of Kansas, Lawrence, Kansas

Larry J. Paulson
Department of Biology and Lake Mead Limnological Research Center, University of Nevada, Las Vegas, Nevada

Kent W. Thornton
Principal Systems Ecologist, FTN Associates, Little Rock, Arkansas

William W. Walker
Environmental Engineer, Concord, Massachusetts

Robert G. Wetzel*
Department of Biology, University of Michigan, Ann Arbor, Michigan

**Present address:* Department of Biology, University of Alabama, Tuscaloosa, Alabama

PREFACE

Our purpose in undertaking the editing of this book revolved around three major factors. First, reservoirs are relatively new limnological systems and have not received extensive limnological or ecological study. While the construction of reservoirs dates to at least 400–300 B.C., most reservoirs in the United States are less than 60 years old. In addition, reservoirs are engineered systems. Scientists tend to study natural phenomena, processes, and systems, whereas engineers concentrate on the design and control of processes or systems. Multi-use demands on reservoirs for municipal drinking water, industrial water supply, irrigation, navigation, hydroelectric power generation, recreation, and aesthetics, however, dictate that the scientific community devote more attention to reservoirs, both for a better understanding of reservoir functioning and for sound ecological management. As illustrated throughout the text, the zonation that occurs in reservoirs has significant management implications for reservoirs.

Second, reservoir research has been too provincial in the past. This has been due, in part, to the belief that each reservoir is a unique system with little potential for transfer of understanding to other systems. While there is no disagreement that any system — lake or reservoir — will have unique attributes, there are certainly general paradigms that can be developed for reservoirs. Some of the differences between lakes and reservoirs, as well as some of their general attributes, are described throughout the text and summarized in Robert Wetzel's chapter.

Finally, reservoirs are fascinating, challenging, and exciting systems of study. Reservoirs are a series of two or three highly interactive ecological systems ranging from riverine ecosystems in the headwater area of the reservoir to the lacustrine ecosystem near the dam. Bruce Kimmel refers to reservoirs as hybrid systems. Reservoirs, like many hybrid systems, exhibit hybrid vigor. Understanding reservoir limnology requires interdisciplinary study and interaction. Reservoirs can serve as laboratories for interdisciplinary studies among hydraulic, hydrologic, and hydrodynamic engineers, microbial, benthic, plankton, and fishery scientists, and aquatic ecologists. Environmental problems require interdisciplinary solutions; no single discipline has the expertise to address all facets of these problems. Traditionally, our pedagogic approach has been to define the problem narrowly so that it could be addressed by students within a single discipline or subdiscipline. These students, then, are typically sent into the marketplace to work as members of teams or organizations — positions which are foreign to them by their training and focus. Certainly, reservoir studies also could be narrowly focused. The perspectives provided in this book, however, seek to expand these interdisciplinary interactions.

The book is intended to be used not as a text but rather as an aid to discussion among system specialists (e.g., lake limnologists) and those in different disciplines through seminars, colloquia, and workshops, and as a focus for a continuing dialogue among lake and reservoir limnologists and engineers.

Kent W. Thornton
Bruce L. Kimmel
Forrest E. Payne

Little Rock, Arkansas
Oak Ridge, Tennessee

CONTENTS

CHAPTER 1

Perspectives on Reservoir Limnology

KENT W. THORNTON

Much of our knowledge of limnology is founded on classic lake studies. For example, Lake Mendota, Linsey Pond, Lake Windermere, and Lawrence Lake are well known throughout the limnological world. The structure, function, and responses of lentic systems are described in *Treatise on Limnology* (Hutchinson 1957, 1967), *Fundamentals of Limnology* (Ruttner 1963), and *Limnology* (Wetzel 1982). Reservoirs have generally been considered synonymous with lakes and are classified as lake type 73 (Hutchinson 1957). The approach to many reservoir limnological studies is identical to the approach used in many lake studies, and the response exhibited by a reservoir is interpreted within the context of conventional wisdom of lake limnology.

Since lakes are created over geologic time and reservoirs are created by man's design, it is important to review the hypothesis that

$$H_o\text{: Lakes} = \text{Reservoirs}$$

Processes such as internal mixing, gas exchange across the air-water interface, redox reactions, nutrient uptake, predator-prey interactions, primary production, and community respiration obviously occur in both lakes and reservoirs. However, because the magnitude and phasing of the forcing functions or driving variables for lakes and reservoirs may not be identical, the responses of the systems may be different. Proper management of our future water resources requires that we review the responses of lakes and reservoirs, speculate on potential differences, and develop hypotheses to explain these potential differences.

PATTERNS

The study of reservoirs and reservoir responses is a study of geologic, macroscale, and mesoscale patterns. Geologic patterns include differences in geology, climatology, and geography. Macroscale patterns occur at the river basin level and include differences in basin hydrology, topography, and weather patterns. Mesoscale patterns occur at the individual reservoir level and are influenced by hydrology, morphometry, and meteorology. These gradients form a highly interactive hierarchy, which cascades from the geologic level to the mesoscale level and influences the limnological response of a reservoir.

Geologic Patterns

The distribution of lakes and reservoirs in the United States significantly influences the limnological responses of these systems. A distributional comparison of natural lakes and Corps of Engineers (CE) reservoirs indicates a bimodal distribution of lakes and, essentially, a unimodal distribution of reservoirs with a reservoir maxima at the lake minima (Figure 1.1). The majority of lakes occurs in the glaciated portion of the United States with the secondary mode representing the Florida solution lakes. The majority of reservoirs are located throughout the southeastern, central, southwestern, and western United States. The small secondary mode in the reservoir distribution at approximately 45° latitude represents the reservoirs on the Missouri and Columbia river system and some smaller New England hydropower projects. Reservoirs are generally constructed for water storage (e.g., flood control, water supply, irrigation, hydroelectric power generation) or where natural water supply is not adequate; that is, reservoirs are constructed where lakes do not exist or are not plentiful.

Geologic differences also influence lake and reservoir water quality. Total dissolved solids (TDS) concentrations are considerably higher in the Great Plains and western streams than in the eastern streams of the United States (Figure 1.2). The distribution of stream TDS overlaps the distribution of many reservoirs, and many reservoirs consequently have high TDS concentrations.

In general, streams located in the geographic areas where natural lakes predominate have much lower suspended solids (SS) concentrations than streams and rivers in major reservoir zones (Figure 1.3). This has signi-

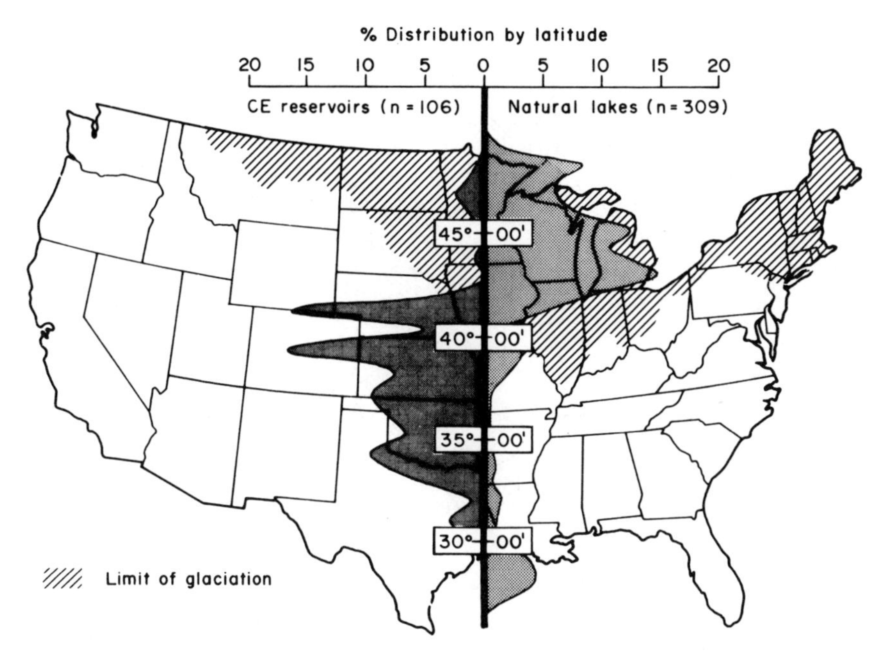

Figure 1.1 Location of lakes and reservoirs (after Walker 1980).

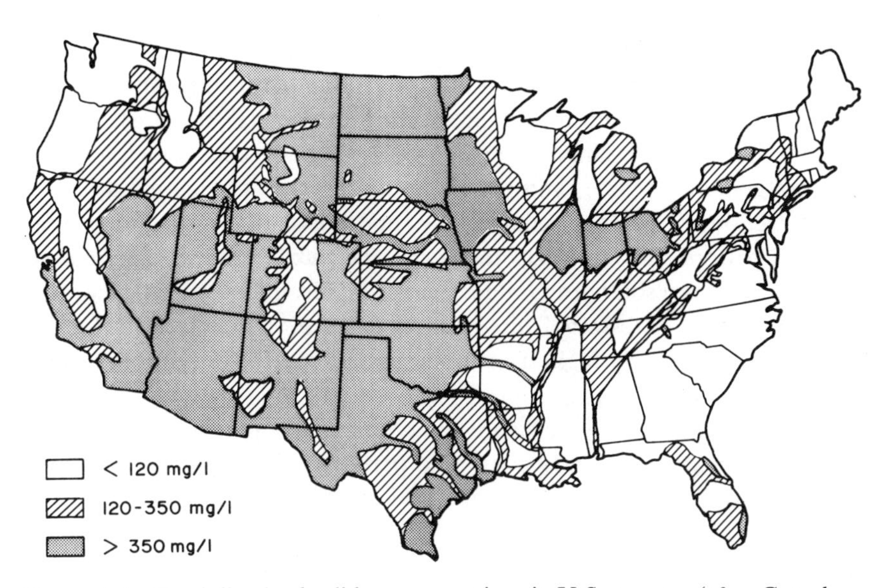

Figure 1.2 Total dissolved solids concentrations in U.S. streams (after Geraghty et al. 1973).

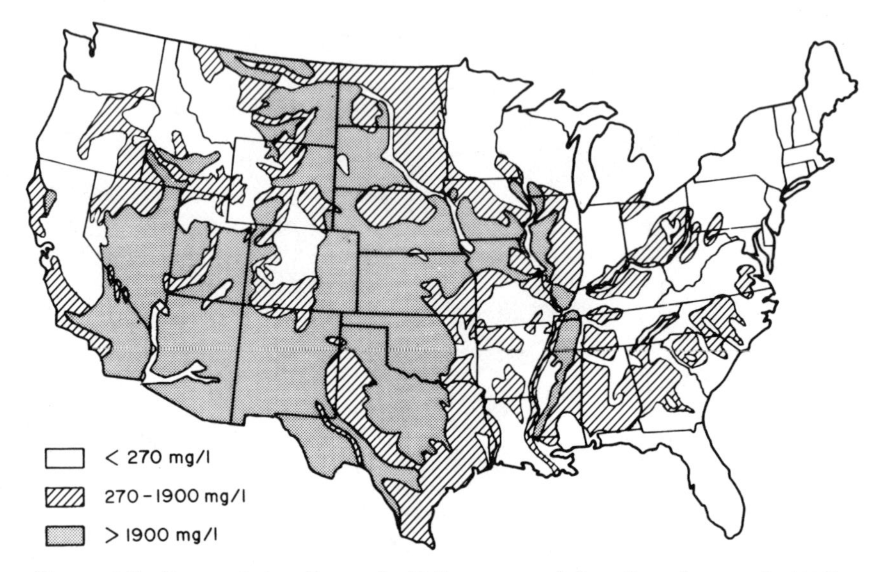

Figure 1.3 Suspended sediment in U.S. streams (after Geraghty et al. 1973).

ficant implications for light penetration and water clarity, nutrient and contaminant transport, and productivity.

Climatological differences also result in patterns. Precipitation-evaporation interactions result in two distinct areas in the United States. In the eastern half of the United States precipitation exceeds evaporation, water is generally plentiful, and lakes are prevalent (Figure 1.4). In the western half of the United States evaporation exceeds precipitation, water is generally scarce, and reservoirs are prevalent (Figure 1.4).

Interactions of geology, climatology, and population distribution also result in patterns in water use. Industrial and public water use is highest in the eastern and southern United States while irrigation is highest in the western and southwestern United States (Figures 1.5 and 1.6). Hydropower use is generally highest in the northeast, northwest, and the Tennessee Valley (Figure 1.7).

The distribution of lakes and reservoirs in conjunction with these geologic, climatologic, and geographic patterns implies potential differences in the limnological responses of lakes and reservoirs.

Macroscale Patterns

Nested within the geologic level are patterns occurring at the drainage basin level. Heterogeneities exist in geology, soil associations, land-use,

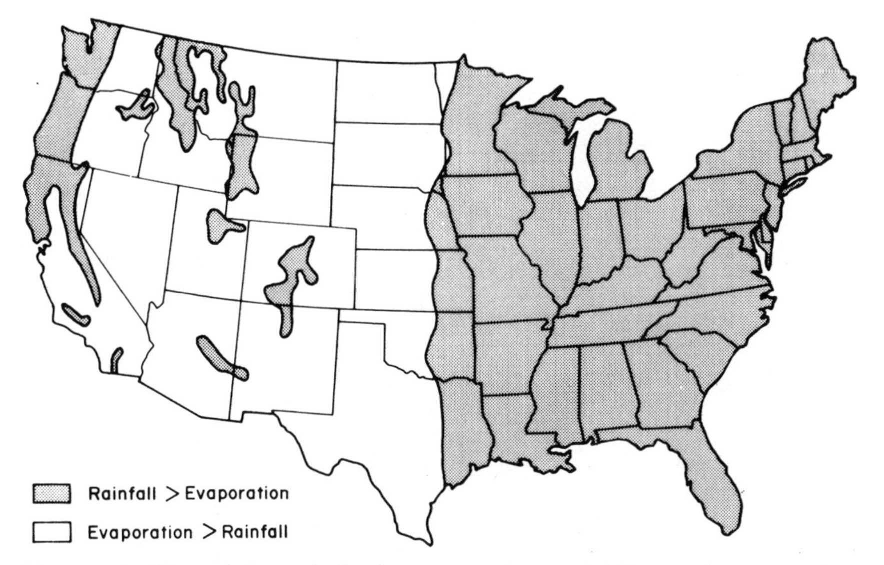

Figure 1.4 Water balance indicating zones where rainfall exceeds evaporation and vice versa.

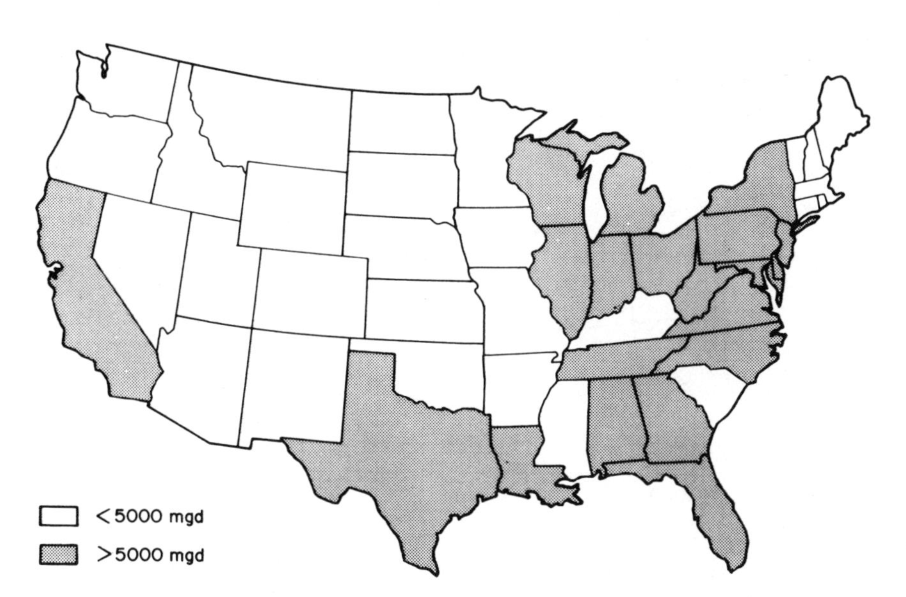

Figure 1.5 Public and industrial water withdrawal in U.S. (after Geraghty et al. 1973).

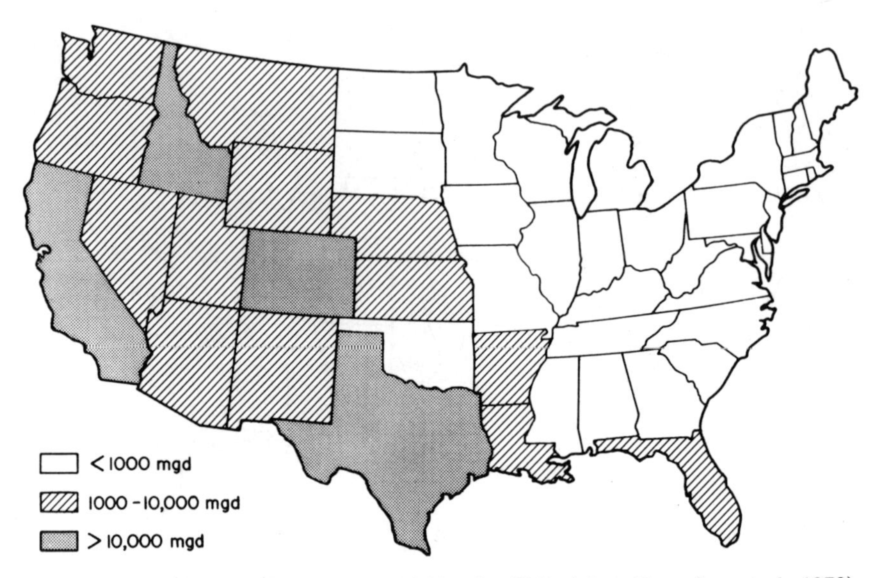

Figure 1.6 Irrigation withdrawals within the U.S. (after Geraghty et al. 1973)

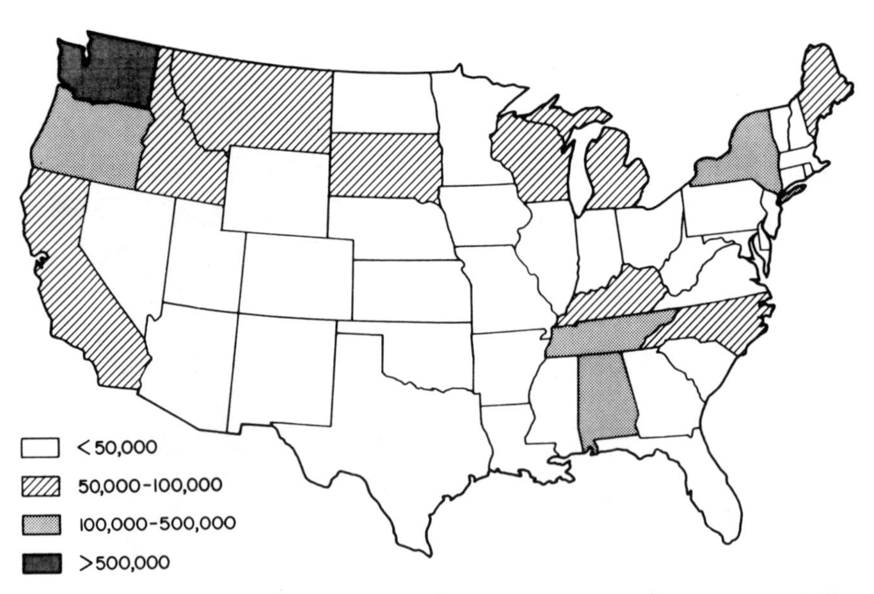

Figure 1.7 Hydropower releases within the U.S. (after Geraghty et al. 1973)

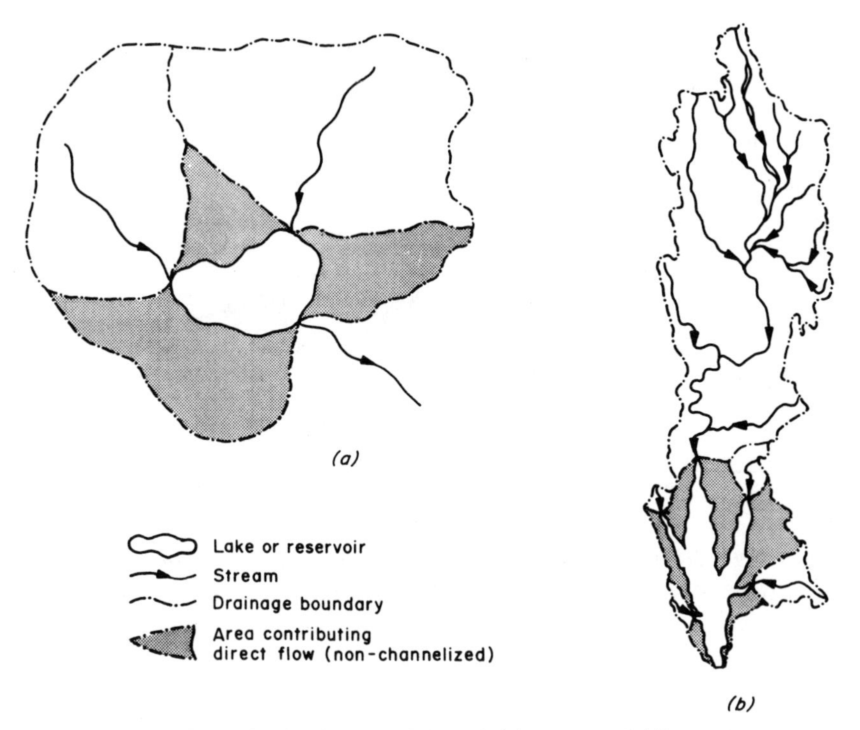

Figure 1.8 Drainage basin characteristics of (*a*) lakes and (*b*) small reservoirs.

weather patterns, and hydrology that influence reservoir limnology. Within a drainage basin these same factors also influence lakes, so why should the responses of these two systems be different?

Many streams and several major rivers originate from headwater lakes, a characteristic of many glaciated lakes. The Mississippi River, for example, originates from Lake Itasca in Minnesota. These lakes generally have a relatively circular drainage basin and a relatively equitable distribution of inflow around the perimeter of the lake (Figure 1.8*a*). Reservoirs, however, are generally located near the mouth or base of the drainage basin. At a minimum, reservoirs are located a considerable distance below the headwater of the stream. The headwater of the reservoir, therefore, serves as a focal point of the basin (Figure 1.8*b*). Reservoirs may receive only a small proportion of their total inflow as direct runoff from the adjacent watershed, with the majority of the water, nutrient, and sediment load entering from one or two major tributaries located a considerable distance from the dam.

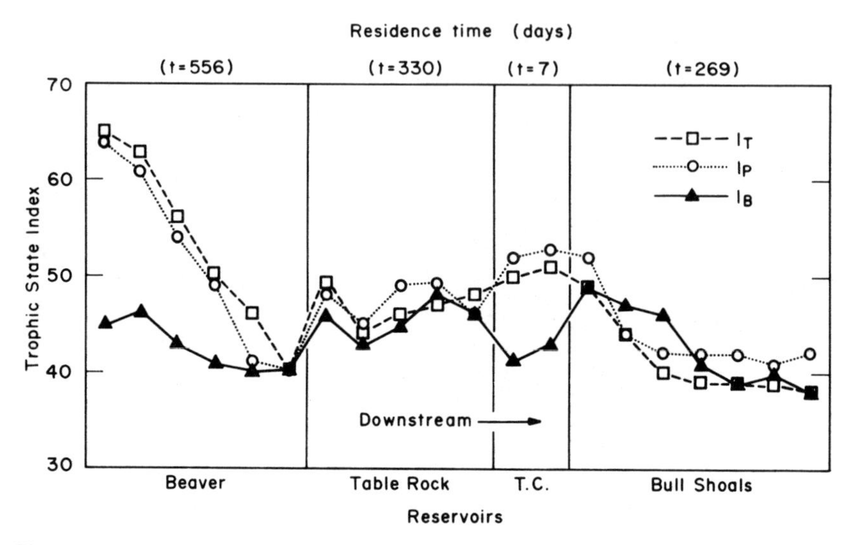

Figure 1.9 Trophic state indices plotted for multiple reservoirs in the White River Basin, Arkansas and Missouri. I_T, I_P, and I_B are trophic state indices based upon Secchi disc transparency, total phosphorus, and chlorophyll-*a*, respectively. T. C. is Lake Tanneycomo (after Walker 1981).

Multiple reservoirs along a river basin also produce unique gradients in reservoir water quality (Pickett and Harvey 1988). The Colorado, Columbia, Missouri, Tennessee, and White River systems are examples of river basins with multiple reservoirs in series. Walker (1981) expressed surface total phosphorus, Secchi depth, and chlorophyll-*a* values in terms of Carlson's (1977) Trophic State Indices (I_p, I_T, and I_B, respectively) for four reservoirs (Beaver, Table Rock, Tanneycomo, and Bull Shoals Lakes) in the White River system (Figure 1.9). The steady decrease in I_p and I_T values from the headwater to the dam in Beaver Lake reflect sedimentation, and the three indices do not converge until the dam. The indices agree reasonably well in the downstream systems. The increase in all indices in the headwater of Table Rock Lake resulted from the contributions of a major point source. With the exception of Lake Tanneycomo, all three projects are deep hydropower reservoirs that thermally stratify. Decreased chlorophyll-*a* indices in Lake Tanneycomo may reflect the effects of cold water releases from the Table Rock hypolimnion because the surface water temperature in Tanneycomo roughly matched the release temperatures (~15°C) from Table Rock Lake (Walker 1981). Lake Tanneycomo's residence time of seven days probably did not permit sufficient time for warming and phytoplankton acclimation (Walker 1981). The indices in Bull Shoals reflect sedimentation.

The Colorado River reservoirs exhibit responses similar to those of the White River. Sedimentation rates, phosphorus concentrations, and productivity are high in the upper end of Lake Powell and decrease downstream to the dam (Gloss et al. 1980). The construction of Lake Powell upstream from Lake Mead has significantly reduced the phosphorus load entering the upper basin of Lake Mead (Paulson and Baker 1981). Lake Mohave, located immediately downstream from Lake Mead, receives hypolimnetic discharges from Lake Mead that are consistently between 11 and 14°C and proceed as an underflow through Lake Mohave during most of the year (Priscu et al. 1981). This cold underflow significantly reduced the hypolimnetic oxygen deficit in the upper end of Lake Mohave (Priscu et al. 1981).

Sediment, nutrient, and temperature gradients are probably only a few of the gradients that exist in multireservoir systems. Limnological studies at the river basin level are rare, and much more research is required to define and understand the hydrodynamic, chemical, and biological interactions that occur through multireservoir operation.

Mesoscale Patterns

Nested within the geological and macroscale levels are patterns occurring at the individual reservoir level. Most of the information on reservoirs, as well as on most aquatic systems, is available on individual systems. While the remainder of this book will discuss specific aspects of reservoir limnology, longitudinal patterns in reservoirs provide the central focus for limnological processes and reservoir system responses (Kennedy et al. 1982, 1985).

An heuristic model is used to describe the development of longitudinal patterns in reservoirs. Occurring along a continuum from river inflow to dam these longitudinal gradients result in the establishment of three distinct zones possessing unique physical, chemical, and biological properties (Thornton et al. 1981). These three zones are a riverine zone, a zone of transition, and a lacustrine zone (Figure 1.10).

The riverine zone is relatively narrow and well mixed, and although velocities are decreasing, advective forces are still sufficient to transport significant quantities of finer suspended particles, such as silts, clays, and

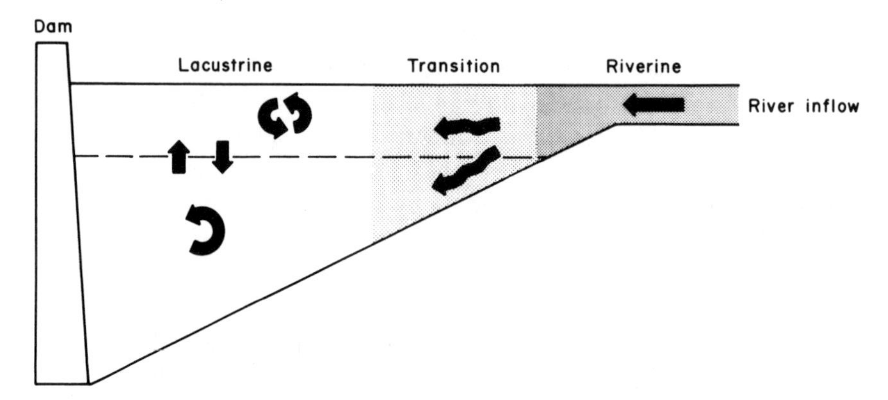

Figure 1.10 Three distinct zones resulting from gradients in reservoirs.

organic particulates (Gordon and Bekel 1985). Light penetration is minimal and generally limits primary production. An aerobic environment is maintained because the riverine zone is generally shallow and well mixed, even though the degradation of allochthonous organic loadings often creates a significant oxygen demand.

Significant sedimentation occurs through the transition zone with a subsequent increase in light penetration (Kennedy et al. 1982). Light penetration may increase gradually or abruptly, depending on the flow regime. At some point within the mixed layer of the zone of transition a compensation point between the production and processing of organic matter should be reached. Beyond this point autochthonous production of organic matter within the mixed layer should begin to dominate.

The lacustrine zone is characteristic of a lake system (Figure 1.10). Sedimentation of inorganic particulates is low, light penetration is sufficient to promote primary production with the potential for nutrient limitation, and production of organic matter exceeds processing within the mixed layer. Entrainment of metalimnetic and hypolimnetic water, particulates, and nutrients may occur through internal seiches or wind mixing during the passage of large weather fronts. Hypolimnetic mixing may be more extensive in reservoirs because of hypolimnetic or bottom withdrawal. Bottom withdrawal removes hypolimnetic water and nutrients and may promote movement of inter- or underflows into the hypolimnion. These mesoscale patterns, however, are emergent properties of an embedded microscale level of gradients.

Table 1.1 Characteristics of reservoirs and glacial lakes[a]

Factor	Characteristic	
	Reservoir	Glacial lake
Shoreline	Astatic	Stable
Water level	Large, irregular fluctuations	Natural
Flushing rate	High	Low
Thermal stratification	Irregular	Natural regime, dimictic
Ionic composition	Variable	Relatively predictable
Sedimentation rate	High	Low
Turbidity	High	Low
Water release level	Variable	Surface
Organic accumulation	Rapid	Slow
Principal source of nutrients	Allochthonous, then autochthonous	Autochthonous
Growth selection	Rapid (r)	Homeostatic (K)
Immigration-extinction	Rapid	Slow

[a]Modified from Ryder 1978.

Microscale

There is a fourth level in limnological studies — the level of measurement or microscale. Processes are similar in both lakes and reservoirs — that is, convective mixing, redox reactions, primary productivity, and predator-prey interactions — and similar measurement techniques are used in both systems, for instance, thermister strings, oxygen probes, and ^{14}C incubations. The lake, reservoir, or river system characteristics become masked when various constituents such as temperature and DO are being measured in the water column. Without knowing which system is being studied, for example, it may be virtually impossible to distinguish a table of DO vs. depth measurements in a shallow lake from those obtained from a river. Interpreting these data, understanding the system response, and placing it in the proper context and perspective, however, occur at the mesoscale level, or one scale above the level of measurement. The interpretation of the DO vs. depth data might be very different with the knowledge that it is a river or lake. Similarly, to understand why a lake or reservoir responds to various hydrometeorological and limnological forces requires a knowledge of that particular system's characteristics. Lakes and reservoirs are not necessarily synonymous aquatic ecosystems. Although some lakes and reservoirs may respond similarly, it is important to consider the particular system characteristics when interpreting system response. As indicated by Baxter (1977, 1985), Goldman and Kimmel (1978), Kennedy et al. (1985), and Thornton et al. (1981), reservoirs may have their own unique responses to forcing function. Some of these responses have been summarized by Ryder (1978) and are listed in Table 1.1.

The purpose of this book is not only to discuss and contrast lakes and reservoirs but also to speculate on comparisons that can be tested (Goldman and Kimmel 1978, Rigler 1975). The following chapters will attempt to show that a reservoir is a "distinct category of tenuous ecosystem retaining specific characteristics of both the lotic and lentic environments" (Ryder 1978, p. 1570).

REFERENCES

Baxter, R. M. 1977. Environmental effects of dams and impoundments. *Ann. Rev. Ecol. and Syst.* 8:255–283.

Baxter, R. M. 1985. Environmental effects of reservoirs. pp. 1–26. In D. Gunnison, ed. Microbial processes in reservoirs. Dr. W. Junk Publishers, Boston, MA.

Carlson, R. E. 1977, A trophic state index for lakes. *Limnol. and Oceanogr.* 22:361–369.

Geraghty, J. J., D. W. Miller, F. van der Leeden, and F. L. Troise. 1973. Water Atlas of the United States. A Water Information Center Publication. Port Washington, NY. 122 p.

Gloss, S. P., L. M. Mayer, and D. E. Kidd. 1980. Advective control of nutrient dynamics in the epilimnion of a large reservoir. *Limnol. and Oceanogr.* 25:219–228.

Goldman, C. R. and B. Kimmel. 1978. Biological processes associated with suspended sediments and detritus in lakes and reservoirs. In J. Cairns, E. F. Benfield, and J. R. Webster, eds. Current perspectives on river reservoir ecosystems. North American Benthological Society Publication No. 2.

Gordon, J. A. and R. M. Behel II. 1985. Suspended sediment characteristics of Lake Cumberland, Kentucky. Pages 259–264. In Proc. N. Am. Lake Mgt. Soc., 1984.

Hutchinson, G. E. 1957. A treatise on limnology: Vol. 1. Geography, physics, and chemistry. John Wiley and Sons, Inc., New York, NY. 1015 pp.

Hutchinson, G. E. 1967. A treatise on limnology: Vol. 2. Introduction to lake biology and limnoplankton. John Wiley and Sons, Inc., New York, NY. 1115 pp.

Kennedy, R. H., K. W. Thornton, and R. C. Gunkel. 1982. The establishment of water quality gradients in reservoirs. *Can. Wat. Resour. J.* 7:71–87.

Kennedy, R. H., K. W. Thornton, and D. E. Ford. 1985. Characterization of the reservoir ecosystem. Pages 27–38. In D. Gunnison, ed. Microbial Processes in Reservoirs. Dr. W. Junk Publishers, Boston, MA.

Paulson, L. J. and J. R. Baker. 1981. Nutrient interactions among reservoirs on the Colorado River. Pages 1647–1656 in H. G. Stefen, ed. Proceedings of the symposium on surface water impoundments. Amer. Soc. Civil Engr., New York, NY.

Pickett, J. R. and R. M. Harvey. 1988. Water quality gradients in the Santer-Cooper Lakes, South Carolina. *Lake and Reser. Mgt.* 4:11–20.

Priscu, J. C., J. Verdium, and J. E. Deacon. 1981. The fate of biogenic suspensoids in a desert reservoir. Pages 1657–1667 in H. G. Stefen, ed. Proceedings of the symposium on surface water impoundments. Amer. Soc. Civil Engr., New York, NY.

Rigler, F. H. 1975. The concept of energy flow and nutrient flow between trophic levels. Pages 15–26 in W. H. Von Dobben and R. H. Lowe-McConnel, eds. Unifying concepts in ecology.

Ruttner, F. 1963. Fundamentals of Limnology. 3rd Ed. (Translat. D. G. Frey and F. E. J. Fry). University of Toronto Press, Toronto, Canada. 295 pp.

Ryder, R. A. 1978. Ecological heterogenity between north-temperate reservoirs and glacial lake systems due to differing succession rates and cultural uses. *Verh. Int. Verein. Limnol.* 20:1568–1574.

Thornton, K. W., R. H. Kennedy, J. H. Carrol, W. W. Walker, R. C. Gunkel, and S. Ashby. 1981. Reservoir sedimentation and water quality — A heuristic model. Pages 654–661 in H. G. Stefen, ed. Proceedings of the symposium on surface water impoundments. Amer. Soc. Civil Engr., New York, NY.

Walker, W. W., Jr. 1981. Empirical methods for predicting eutrophication in impoundments. Phase I: Data base development. Technical Report E-81-9, prepared by William W. Walker, Jr., Environmental Engineer, Concord, MA, for the United States Army Engineer Waterways Experiment Station, CE, Vicksburg, MS.

Wetzel, R. G. 1983. Limnology. Saunders College Publishing, Philadelphia, PA. 767 pp.

CHAPTER 2

Reservoir Transport Processes

DENNIS E. FORD

The movement and mixing of dissolved and particulate matter within a water body results from a number of complex, highly interdependent, physical transport mechanisms. Because these mechanisms influence the environment (i.e., temperature, light, and chemical regimes) in which aquatic organisms exist, an understanding of reservoir transport processes is essential to understanding reservoir limnology.

The objective of this chapter is to introduce reservoir transport mechanisms that impact reservoir water quality and distinguish reservoir limnology from classical limnology. These mechanisms include advection, convection, diffusion, dispersion, entrainment, mixing, settling, and shear. The relative magnitude and importance of these individual mechanisms vary from reservoir to reservoir and depend on reservoir morphometry, density stratification, meteorological forcing, inflow, outflows, and project operation. The time scales for these mechanisms vary from fractions of a second for turbulence to months for basin-wide mean motion (i.e., a seasonal time scale), and the length scales of motion vary from fractions of a centimeter to the size of the basin. For in-depth reviews and specific details the reader is referred to Boyce (1974), Mortimer (1974), and Csanady (1975) on dynamics of large lakes, Hansen (1978), Fischer et al. (1979), Imberger and Hamblin (1982), and Ford and Johnson (1986) on mixing dynamics, Tennekes and Lumley (1972) on turbulence, Ford and Johnson (1983), Imberger (1987), Johnson and Ford (1987) on inflow dynamics, and Imberger (1980), Smith et al, (1987), and Monismith et al. (1988) on outflow dynamics and selective withdrawal.

In this chapter the major transport mechanisms will first be defined, then the annual temperature stratification cycle will be described in terms of potential energy. Finally, the major factors influencing the potential energy of stratification (i.e., reservoir morphometry, meteorological forcing, inflows, outflows, and project operation) will be discussed with respect to their impact on mixing and transport in reservoirs.

Transport Mechanisms

Following the conventions of Fischer et al. (1979) and Ford and Johnson (1986), the major transport mechanisms in reservoirs and lakes are:

Advection. Advection is transport by an imposed current system. In reservoirs advection may be caused by river inflows, outflows, and wind shear at the air-water interface (Figure 2.1).

Convection. Convection is vertical transport induced by density instabilities. When the surface layer of a reservoir cools, the resulting denser surface water sinks, generating convection motions (Figure 2.1).

Turbulence. Turbulence is sometimes described as a family of eddies (i.e., rotating regions of fluid) that range in size from the physical limits of the flowfield down to molecular motion. The disadvantage of this portrayal is that it is difficult to separate wave motions from turbulence. More precisely, turbulent flows are irregular (random), diffusive (produce mixing), rotational (overturning motions), time-varying, and dissipative (decay rapidly without a continual source of energy) (Tennekes and Lumley 1972). In reservoirs turbulence can be generated by the wind, inflows, outflows, convection, boundaries, etc. (Figure 2.1).

Diffusion. Diffusion is a mechanism where differences in mean concentrations are always reduced. Molecular diffusion is a mechanism in which a certain property of a fluid is transferred down a concentration gradient by the random motion of molecules without any overall transport of a fluid taking place. Molecular diffusion is an intrinsic property of the fluid and varies from solute to solute. In contrast, turbulent diffusion is the random scattering of particles by turbulent motion. Turbulent diffusion is usually considered analogous to molecular diffusion but with a

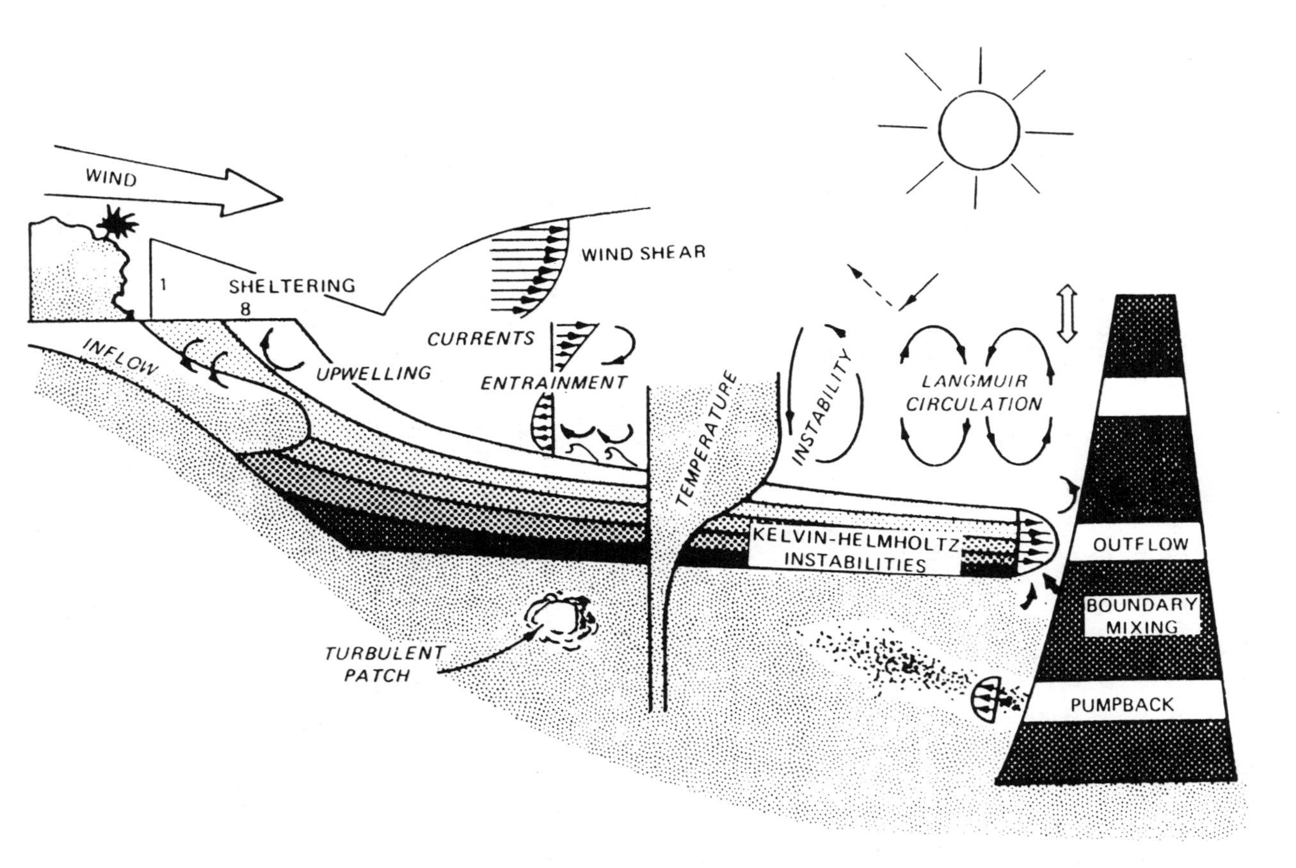

Figure 2.1 Pictorial representation of reservoir mixing processes.

larger "eddy" diffusion coefficient. Turbulent diffusion is *not* a property of the fluid but rather a function of the flow regime.

Shear. Shear is the advection of a fluid at different speeds at different positions. Shear, therefore, requires a velocity gradient. In lakes and reservoirs shear can be generated at the air-water interface by the wind along the bottom boundary, by inflow currents, and internally by density currents (Figure 2.1).

Dispersion. The combined effects of shear and diffusion is called dispersion. Since there is usually little information on the velocity distribution in reservoirs it is difficult to separate dispersion from diffusion. In general, dispersion should dominate in the headwater regions of a reservoir, where the riverine velocities are still large, and diffusion should dominate in the main pool of the reservoir, where current velocities are small (Figure 2.1).

Entrainment. Entrainment is an advective-type transport mechanism where the thickness of the stirred layer grows by entraining (trapping) fluid from the unstirred layer. The interface between the stirred and unstirred layers, therefore, advances into the unstirred layer. In reservoirs, metalimnetic water is entrained into the epilimnion by wind-generated turbulence or convection, and the depth of the epilimnion grows (Figure 2.1). Ambient reservoir water can be entrained into an inflow density by turbulence generated at the sediment-water interface. The thickness of the inflow density will therefore increase. In contrast to diffusion, entrainment sharpens gradients.

Mixing. Mixing is any mechanism or process that causes a parcel of water to blend with or be diluted by another. Mixing includes diffusion, shear, dispersion, and entrainment.

Settling. The sinking of particles with densities different from the surrounding fluid is called settling. Settling is discussed in Chapter 3.

All of these transport mechanisms occur in all reservoirs at one time or another. It is also probable that different transport mechanisms will be dominating at different locations within one reservoir at a specific time. The specific types or type of transport mechanisms that occur will depend on the type, magnitude, duration, and variation of the forcing variables (i.e., energy sources).

Stratification and Potential Energy

The principal factors influencing the formation, strength, and extent of thermal (density) stratification are the density of water, solar radiation, energy transfer at the air-water interface, reservoir morphometry, and the mixing resulting from advection and wind-induced phenomena. Because an understanding of mixing processes in reservoirs has benefited from energy considerations, the principal factors governing stratification will be discussed in terms of kinetic energy (KE) input and changes in the potential energy (PE) of stratification.

The potential energy of a reservoir is defined as:

$$\mathrm{PE} = mgh = \int_{o}^{Z_m} gzA(z)\, \rho\, (z,\, t)\, dz$$

where m is the total mass of the reservoir
g is the acceleration due to gravity
h is the elevation of the center of mass of the reservoir
z is the elevation above the reservoir bottom
$A(z)$ is the plan area of the reservoir at elevation z
$\rho\,(z,\,t)$ is the reservoir density at elevation z and time t
Z_m is the maximum elevation

The PE of a reservoir can be changed by adding heat to modify the water density (i.e., mass) and/or by changing the elevation of the center of mass. For a well-mixed body of water the center of mass is the center of volume. For a stratified body of water the surface waters are less dense (i.e., have less mass) than the underlying waters, and the center of mass is located deeper than the center of volume. KE is therefore required to mix a stable stratified fluid because the center of mass must be raised. For example, completely mixing the two-layered fluid in Figure 2.2 changes the PE by $\Delta\rho\ VH/8$ and raises the center of mass to $H/2$. The potential energy concept is similar to Birge's wind work and Schmidt's stability calculations (Hutchinson 1957).

To illustrate how the PE and KE interact to form a seasonal temperature cycle, a reservoir with a long residence time will first be used as a simple example (Figure 2.3) to minimize the effects of inflow and outflow. In February the reservoir was isothermal and the entire water column mixed vertically. The center of mass was therefore located at the center of volume. At these low temperatures the density differences due to

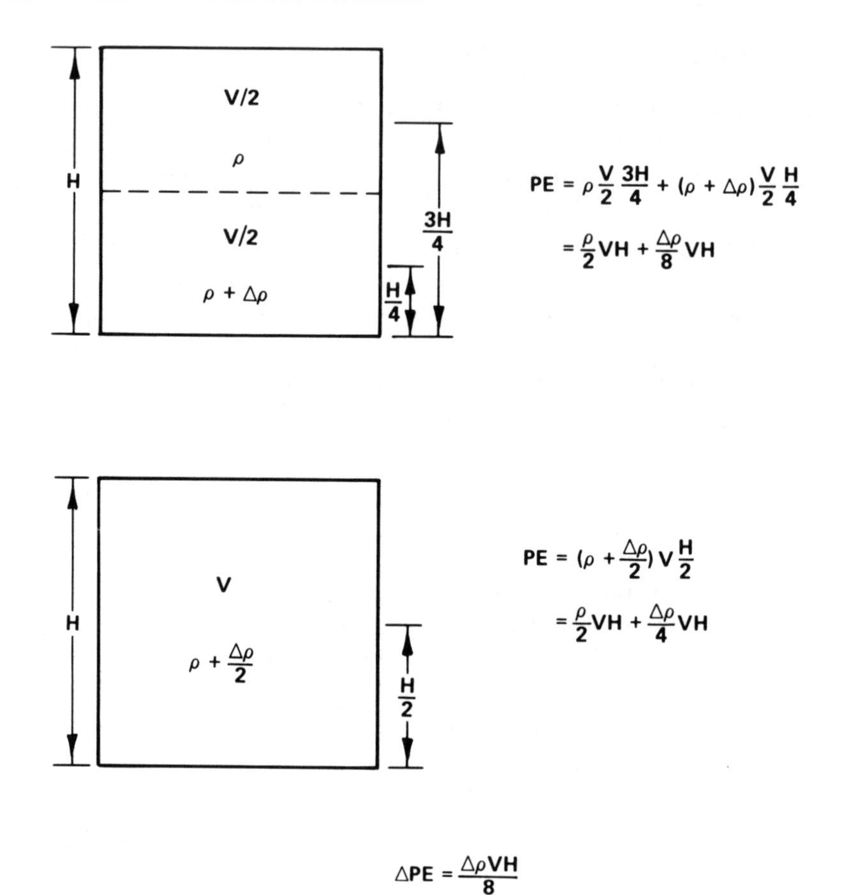

Figure 2.2 Change in potential energy resulting from destratification of a two-layered system.

warming at the water surface were not sufficient to prevent complete mixing. As the water column warmed, density differences due to warming at the water surface increased, and greater energy inputs were required to mix the entire water column. Even though the water column remained well mixed and the location of the center of mass did not change, the PE still increased because the water colunn warmed. In Figure 2.3 stratification started to form near the bottom of the lake because the density differences and resulting PE were small compared to the KE input from the wind (Ford and Stefan 1981). As the solar radiation increased, water temperatures increased, density differences increased, and the thermo-

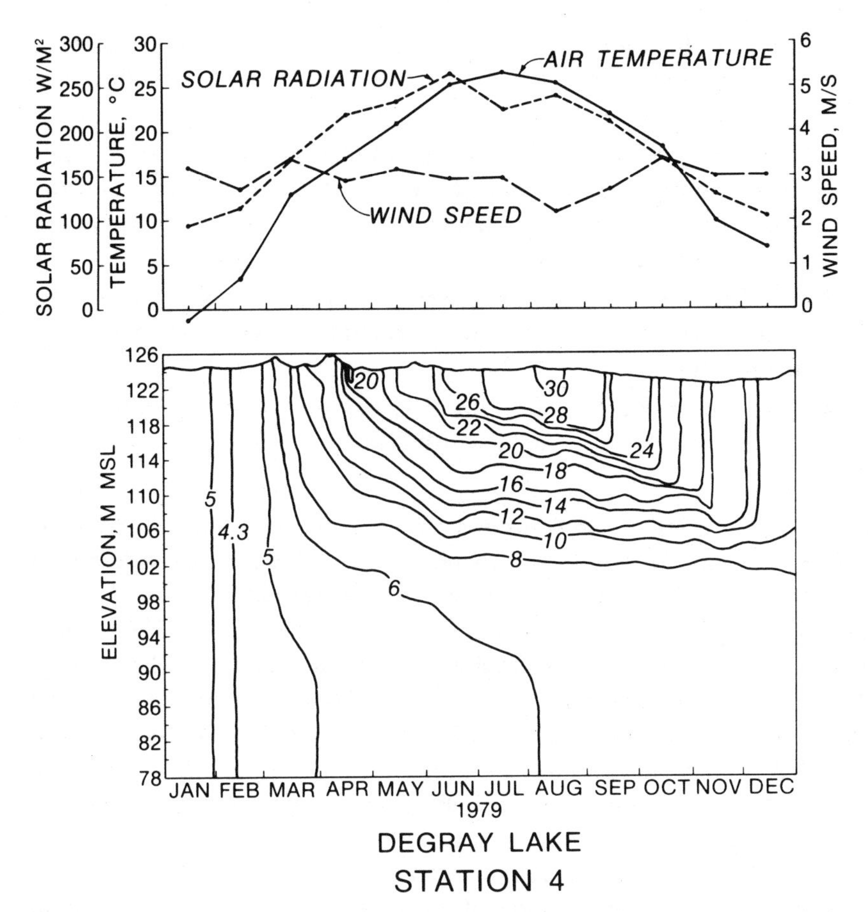

Figure 2.3 Comparison of seasonal temperature isotherm in DeGray Lake, Arkansas, with monthly meteorological conditions.

cline moved upward because the KE input could not overcome the ever increasing PE (i.e., buoyancy forces) and raise the center of mass. The minimum thermocline depth was achieved around the time of summer solstice or maximum heat input (i.e., maximum PE input). Once stratification formed the hypolimnion temperature increased only slightly until fall overturn. Fall overturn occurred when the temperature of the surface strata approached the hypolimnia temperatures.

Four generalizations concerning stratification in natural lakes or reservoirs dominated by meteorological forcing (long residence times) can be made:

1. Because the major factors responsible for stratification (e.g., wind and solar radiation) act at the air-water interface, horizontal variations are minimized.
2. There is not much variability in stratification from year to year. Stratification formation and fall overturn occur at approximately the same time each year. Hypolimnetic temperatures are similar (within 1 or 2°C) from year to year.
3. The larger the surface area, the greater the kinetic energy input, the longer the periods of turnover, and the higher the hypolimnetic temperatures.
4. The deeper the lake, the less KE per unit volume available for mixing, and the stronger the stratification.

In reservoirs inflows and outflows can be significant sources of both KE (advection) and PE (heat) and thereby modify the previously described stratification patterns. Because most of the water entering the reservoir enters through a few major tributaries, the input KE and PE is concentrated at distinct points that promote horizontal variations. The input of KE and PE from inflows is also more irregular when compared with the input of KE from the wind and PE from solar radiation. As shown in Figure 2.3, the wind (i.e., KE input) is relatively constant throughout the year and the solar radiation (i.e., PE input) varies sinusoidally with a maximum in June. This pattern is repeated year after year. In contrast, major inflow events can occur almost anytime (Figure 2.4). The timing and magnitude of a spring inflow event can shift the onset of stratification by several weeks. In natural lakes, where the major source of KE is through the air-water interface (i.e., the wind), thermal stratification prevents the KE from directly entering the hypolimnion and thereby minimizing hypolimnetic mixing. In reservoirs, KE and PE can enter the hypolimnion directly via density currents. This increased level of mixing, coupled with the direct input of PE (heat), causes the hypolimnetic temperatures to increase through the summer (Figure 2.5). Stratification patterns in reservoirs can therefore vary significantly from year to year (Figure 2.5), and generalizations based on morphometry are probably not valid.

Mixing and entrainment are often not gradual, continuous processes in reservoirs, as indicated by the seasonal temperature structure in Figure 2.3. Dynamic processes result from the interactions and variations of

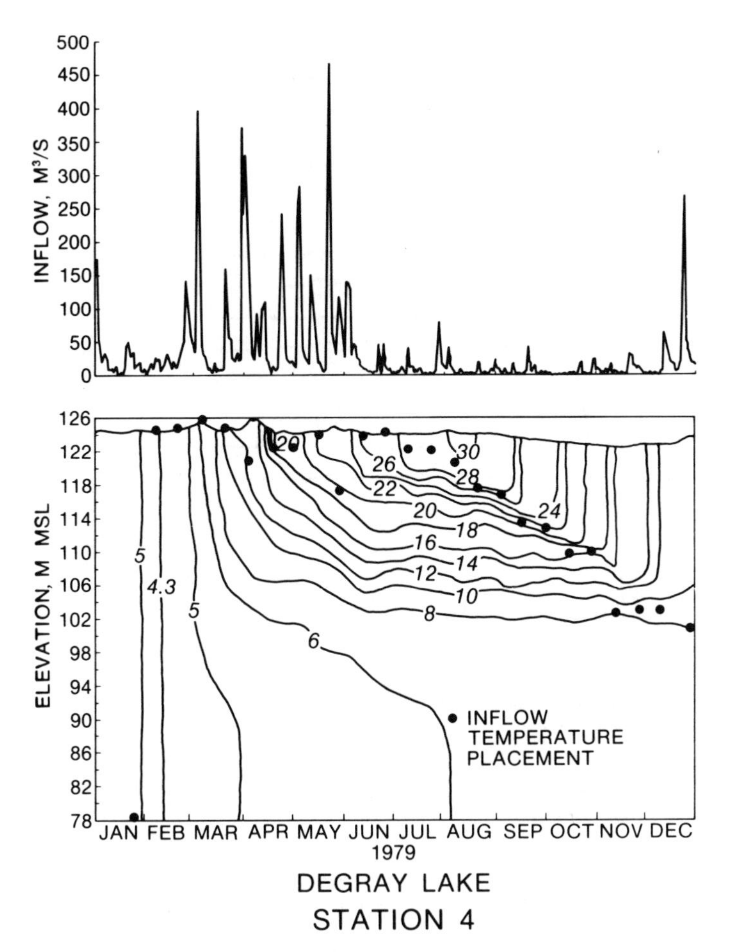

Figure 2.4 Comparison of seasonal isotherms in DeGray Lake, Arkansas, with mean daily inflows.

solar radiation, strong winds, inflows, outflows, hydropower operations, etc. Examples showing the extent of mixing resulting from several hours of strong winds are given in Figure 2.6. Annual entrainment of nutrients from the metalimnion into the euphotic zone can therefore be considered a relatively rapid process rather than a gradual, continuous process.

In addition to the seasonal cycle in temperature there are also diel and synoptic cycles in temperature (Figure 2.7). These cycles result from variations in solar radiation and mixing. Field measurements are usually taken during daylight hours and frequently in midafternoon when diurnal

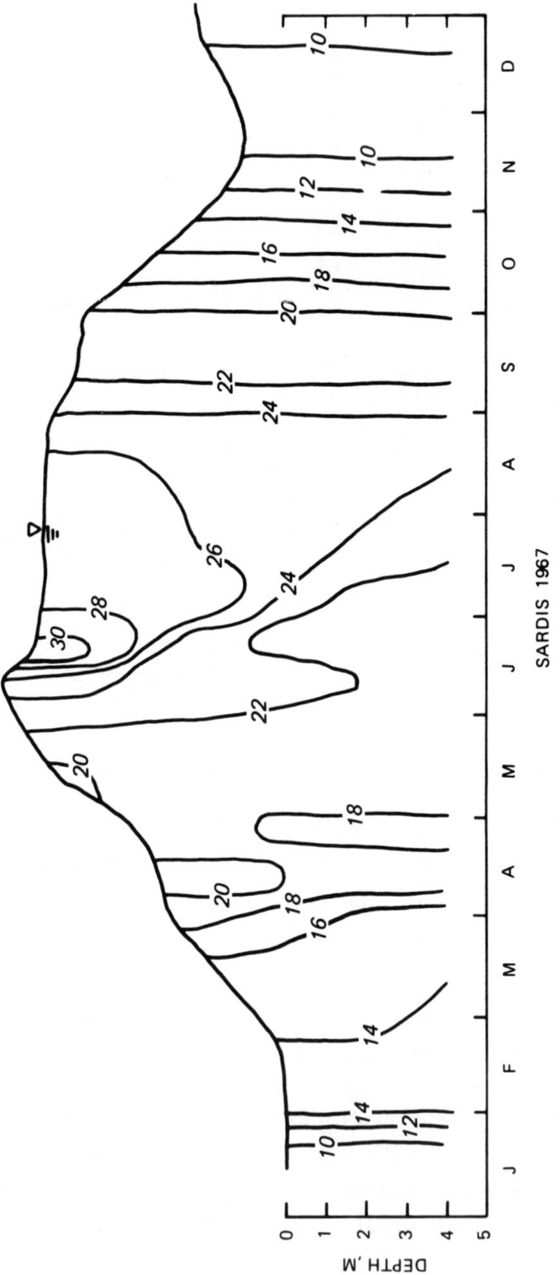

Figure 2.5 Annual isotherm for a reservoir dominated by inflows.

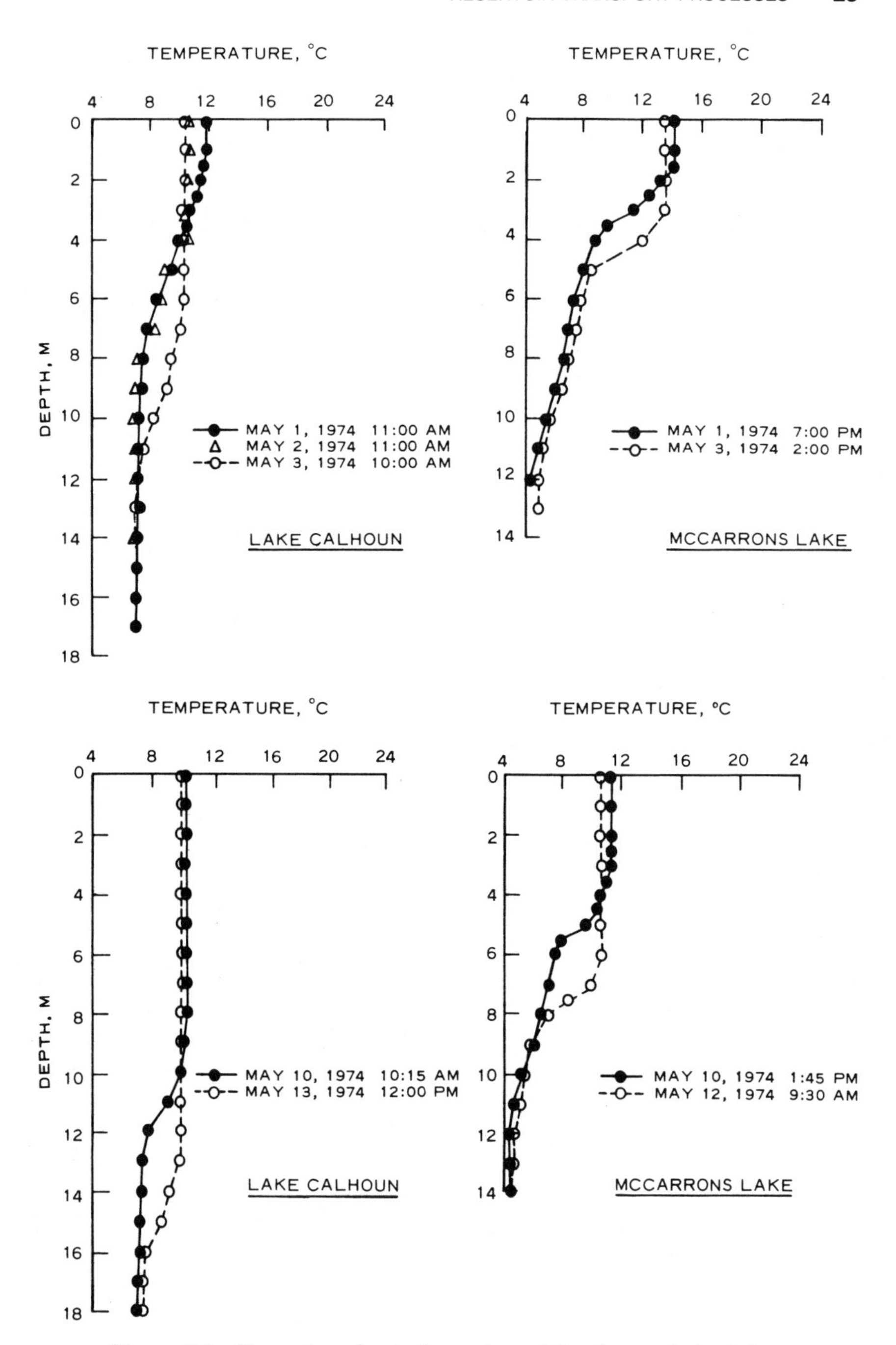

Figure 2.6 Examples of entrainment resulting from wind mixing.

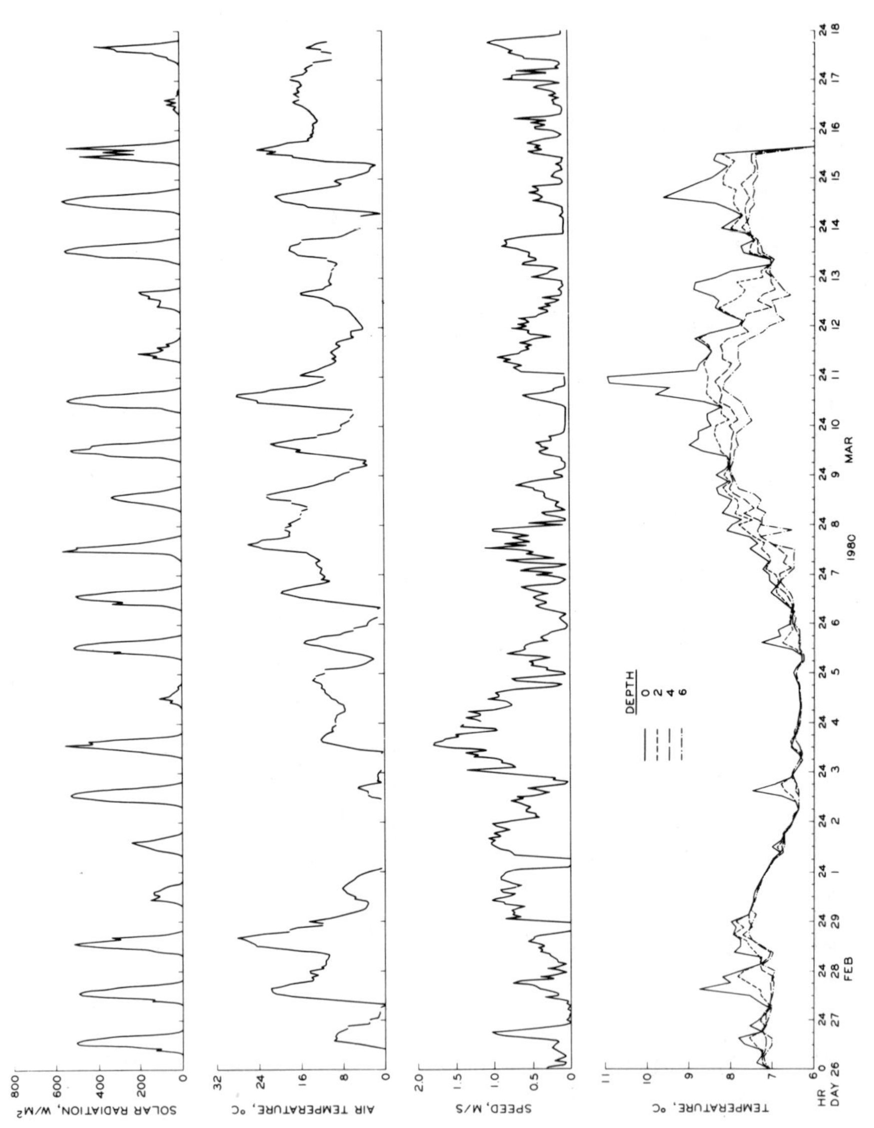

Figure 2.7 Comparison of water temperature in DeGray Lake with meteorological forcing.

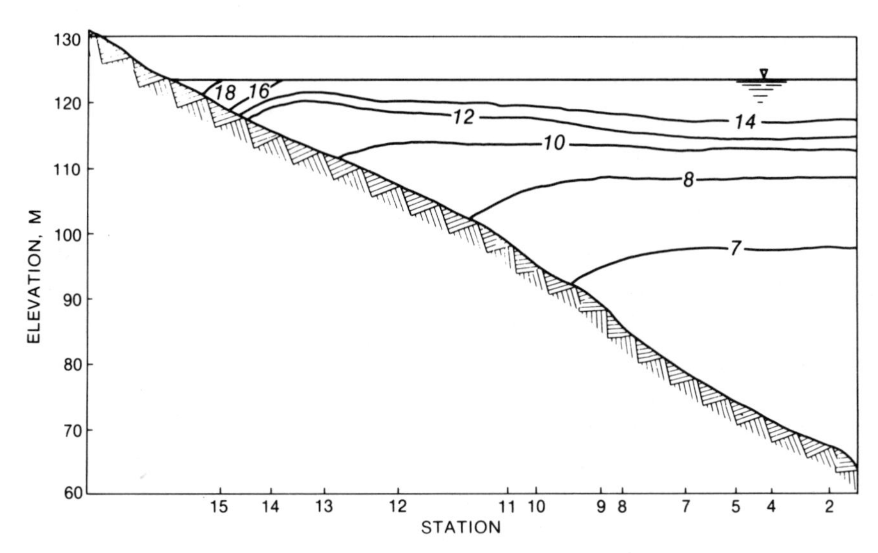

Figure 2.8 An example of horizontal variation in water temperature, DeGray Lake, Arkansas, 2 April 1976.

heating is maximum: diel variations can be as large as 7°C or more but are typically a few degrees centigrade. The magnitude of the variations will depend on the size and depth of the lake, among other factors.

Horizontal variations also occur and result from differential heating, inflow, or mixing. Differential heating takes place when the smaller volume of water in shallow areas, such as littoral zones and headwater regions of the impoundment, warms or cools more rapidly than the water of more open-water regions. In large lakes this phenomenon is significant and results in the formation of thermal bars (Wetzel 1975). Similarly, rivers flowing into a reservoir may be of different temperatures and can create horizontal variations. An example is shown in Figure 2.8. Horizontal variations typically create temperature differences of 1 or 2°C or more.

Horizontal variations in temperature (density) may result in unstable conditions that generate currents (i.e., transport). During periods of warming the warmer surface water in the shallow littoral zones may flow outward along the water surface to the pelagic region of the reservoir. To preserve conservation of mass a return current must exist in the bottom waters. During periods of cooling the reverse is true. This mode of transport between the littoral and pelagic regions is probably not significant except during prolonged periods of heating and cooling (i.e., several days).

Meteorological Forcing

The type and magnitude of meteorologic forcing is dependent on the reservoir location (i.e., latitude, longitude, and elevation) and the surrounding terrain. For the most part meteorological forcing acts at the air-water interface and is therefore dependent on the size and shape of the water surface and on the surrounding terrain. Of all the meteorological factors influencing movement and mixing in reservoirs, solar radiation and wind are probably the most important.

The energy available to warm the waters of a reservoir ultimately comes from solar radiation, which varies seasonally. The seasonal variation of solar radiation follows a sinusoidal curve with a maximum in the northern hemisphere in late June (Figure 2.3). In addition, diurnal cycles also occur. Large variations in solar radiation due to passing clouds are commonplace (Figure 2.7). Water temperatures respond to both of these cycles with a slight delay.

In nearly all lakes and reservoirs wind is the major source of energy (kinetic) for mixing. Mixing results from the interaction and cumulative effects of wind-induced shear at the air-water interface (e.g., currents, surface waves, internal waves (seiches), and entrainment). Like solar radiation, the wind is highly variable with seasonal, synoptic, and diurnal cycles. Synoptic cycles correspond to the passage of major weather systems and commonly have a period of 5–7 days. Like solar radiation, the wind has a diurnal cycle, but the maxima do not necessarily coincide. For example, in Minneapolis, MN, the wind does not normally peak until the middle of the afternoon.

The manner in which a reservoir responds to hydrometeorological forcing depends on the morphometric characteristics of the basin and its location. Reservoirs are typically built in deep valleys after the confluence of several tributaries. The resulting water body is long and highly dendritic (e.g., Figure 2.9). Sheltering of the water surface from the wind by the surrounding terrain may be significant. The complicated geometry will also limit the fetch and complicate circulation patterns.

In summary, the meteorological phenomena responsible for mixing and movement in reservoirs are highly variable. A reservoir is always in a state of flux and is never in equilibrium (i.e., steady state) with the forcing functions.

Inflows

In natural lakes inflows are usually restricted to overland flow and small streams. Their impact is usually confined to the littoral zone and surface

Figure 2.9 Dendritic shape of a reservoir, DeGray Lake, Arkansas.

waters. In contrast, in reservoirs major river inflows usually enter at the upper end of the reservoir. The quantity of runoff will be dependent on the shape and size of the watershed, antecedent conditions, and the distribution of precipitation. In northern latitudes runoff from spring snowmelt may contribute the majority of water to a reservoir. In other locations runoff from individual storm events dominates the water budget. During elevated flow events, for example, inflows may pass through a reservoir in a few days, even though the theoretical hydraulic residence time is much longer (e.g., 0.8 years in DeGray Lake). The mean annual hydraulic residence time, therefore, may not be an appropriate measure of the influence inflows have on reservoir mixing.

Because the inflow density usually differs from the density of the reservoir water surface, inflows enter and move through reservoirs as density currents. Bell (1942) defined a density current as a gravity flow of a liquid or a gas through, under, or over a fluid of approximately equal density. Density flows differ from normal fluid flows because the buoyancy of the surrounding fluid reduces the gravity force by the normalized density difference, $\Delta\rho/\rho$ (i.e., reduced gravity force = $g\Delta\rho/\rho$). Vertical movements such as waves are therefore exaggerated at the density interface within the metalimnion compared to the air-water interface.

Density differences can be caused by temperature, total dissolved solids, and suspended solids. In many reservoirs characterized by low TDS and SS concentrations, density differences are predominantly caused by temperature. For example, at 25°C it takes approximately 330 mg/l of TDS or 420 mg/l of SS (specific gravity = 2.65) to equal the density difference caused by a 1°C temperature change. At temperatures less than 10°C solids concentrations of the order of 30 mg/l can modify the water density. The contribution of SS to water density is complicated because SS settle out of the water column in the headwaters of a reservoir (see Chapter 3). SS concentrations also tend to increase with flow (see Chapter 3). In contrast, TDS concentrations can either increase with increasing flow (a first-flush phenomenon from saline watershed) or decrease with increasing flow (a dilution effect). During storm events Ford and Johnson (1983) showed large, unpredictable variations in inflow water density resulting from variations in temperature, TDS, and SS.

Depending upon the density difference between the inflow and reservoir, density currents can enter the epilimnion, metalimnion, or hypolimnion (Figure 2.10). When the inflow density is less than the water surface density, the inflow will flow over the lake water (i.e., overflow).

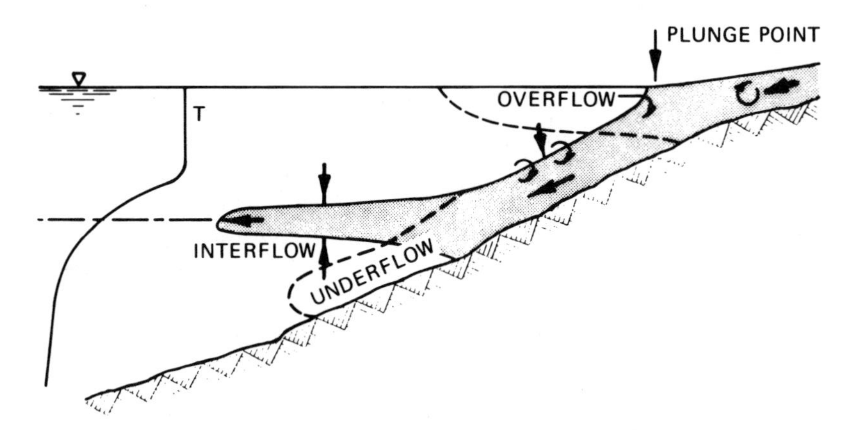

Figure 2.10 Density inflows to impoundments.

This condition typically occurs during spring when the inflowing river water is warmer than the lake water.

Several factors complicate the analysis of overflows. First, since the motive force is the excess hydrostatic pressure, the current will spread in all directions unobstructed by boundaries. Second, the temperature (density) difference is quickly dissipated by heat transfer at the air-water interface. Third, wind shear can direct the overflow into a cove or prevent it from mixing downreservoir. Horizontal dispersion will also be enhanced from wind shear. Fourth, vertical mixing resulting from wind shear and/or convective cooling can distribute the density current through the water column.

If the inflow density is greater than the density of the water surface, the inflow will plunge beneath the water surface (Figure 2.10). The plunge point is sometimes visible because of turbidity of floating debris, indicating a stagnation point or a point of convergence. The location of the plunge point is determined by a balance of the stream momentum (advective force), the pressure gradient across the interface separating the river and reservoir waters (bouyancy force), and the resisting shear forces (wind, bed friction). The position of the plunge point is highly dynamic and changes with both flow and density. During a storm event it can move ten kilometers or more in a single day (Ford and Johnson 1983). The location of the plunge point is more sensitive to changes in flow than to changes in the magnitude of the density difference between the inflow and reservoir surface water (Ford and Johnson 1980).

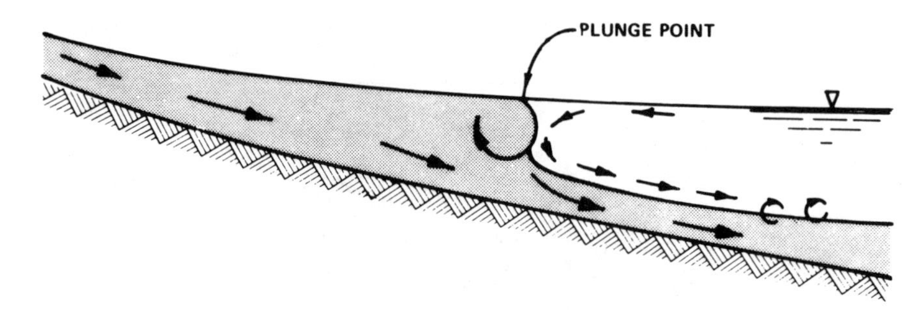

Figure 2.11 Pooling at the plunge point.

Knapp (1942) described the flow in the vicinity of the plunge point not as plunging beneath the surface but as flowing out of the bottom of the mixing zone (Figure 2.11). Flow and material, therefore, have a tendency to pond at the plunge point. Ford et al. (1980) verified this phenomenon using dye studies in DeGray Lake, AR. While some mixing with reservoir waters occurs at the plunge point, it is difficult to quantify. Estimates vary from less than 10% to over 100%, but no consistent theory is currently available to quantify the mixing (Ford and Johnson 1983).

The ponding of inflowing materials at the plunge point may have a significant effect on reservoir water quality. It is not known, however, what happens to these materials as the plunge point advances into the reservoir and retreats back upstream. The dissolved materials probably remain in the surface waters. In Chapter 1 and in Thornton et al. (1980), the concept that a reservoir can be divided into three zones (riverine, transition, and lacrustrine) was presented. The transition zone may actually be the zone where the plunge point occurs. Its upstream boundary is determined by the location of the plunge point at low flow conditions, and its downstream boundary is determined by the location of the plunge point at high flow conditions. The transition zone is the region the plunge point moves back and forth through in response to changing flow conditions.

After the inflow plunges it can follow the old river channel (thalweg) as an underflow. Underflows have been observed to extend to the dam when the inflow is sediment-laden (Grover and Howard 1938). The speed and thickness of the underflow can be determined by assuming a two-layered system and a flow balance between the shear forces and the acceleration due to gravity (Ford and Johnson 1981). Entrainment of reservoir surface waters into an underflow results from turbulence generated by bottom roughness. It is also possible that materials found in the underflow can be entrained into the surface waters by the wind.

The density interflows or intrusion occurs when a density current leaves the river bottom and propagates horizontally into a stratified body of water. Interflows are common in reservoirs and usually occur in mid to late summer, when inflow temperatures are less than the water surface temperature and greater than the hypolimnetic water temperature. Intrusions differ from overflows and underflows because an intrusion moves through a reservoir at a level where the intrusion and reservoir densities are similar. Intrusions require a continuous inflow and/or outflow for movement or they stall and collapse. Ford and Johnson (1981) showed a storm event that stalled in the metalimnion of DeGray lake when the outflow was stopped. The associated constituent load remained in place as a thin lens. Entrainment into an interflow can usually be neglected because the density gradient in the metalimnion creates strong buoyancy forces that inhibit mixing.

Because inflow densities (temperatures) are continuously changing, the level at which an intrusion moves through a reservoir will also change. This is especially true during storm events, when inflow densities change rapidly due to increased flow, temperature changes, and solids load. Ford and Johnson (1983) showed that during one storm event on DeCray Lake, the vertical placement of inflows varied by over five meters. Since inflowing phosphorus concentrations typically load on the rising side of a hydrograph (first-flush phenomenon) and nitrogen loads on the falling side, it is possible that these two constituents could enter a reservoir at different levels.

Interflows have been observed to follow the thalweg of the old river channel and not mix laterally across the reservoir (Ford and Johnson 1981). This movement may in part be due to reservoir morphometry, dendritic shapes of reservoirs, changes in bottom roughness, and/or clearing of timber. It implies, however, that lateral variations in reservoir water quality may also exist.

It is frequently assumed that once an inflow has plunged and formed an underflow or interflow, its constituent load is isolated from the surface waters. Although this assumption has been shown to be true in many situations, recent studies have indicated that mixing from meteorological forcing can entrain the inflow constituents into the surface waters. This entrainment can occur from convection mixing at night (Ford et al. 1980) and from wind-generated phenomena such as internal waves (Carmack

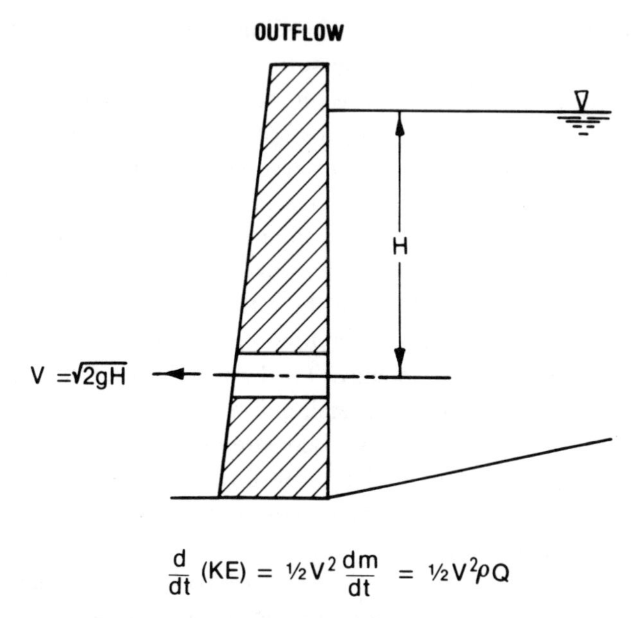

Figure 2.12 Relationship of potential and kinetic energy at an outlet.

and Gray 1982). These mixing processes are the same processes that mix metalimnion waters into the epilimnion of natural lakes.

Outflows and Reservoir Operation

Reservoirs are constructed to stabilize the flow of water. Water is stored during wet periods to be used during periods of low flow or to prevent flood damages downstream. The type of outlet structure and project operation is totally dependent on project purposes.

When water is released from a reservoir, potential energy is converted into kinetic energy and mixing (Figure 2.12). Mixing is restricted to the zone outflow and proportional to the third power of the discharge since $KE = \frac{1}{2}mv^2 \propto Q^3$ and $m = \rho\, Q^t$ and $v = Q/A$.

In small agricultural reservoirs the outlet structure is typically a siphon or closed conduit spillway (see Lindsey and Franzini 1972 for details). These structures automatically maintain the water surface at a specific elevation by releasing surface water over a weir-type structure. This form of surface withdrawal is similar to outflows from natural lakes and has minimal impact on the in-pool mixing.

Large flood control projects also have spillways to discharge large floods and prevent damage to the dam. These spillways may be controlled (with crest gates to control the release rate) or uncontrolled. In either case, because the majority of the water storage is located below the spillway crest elevation, the project is also equipped with a sluiceway or conduit to release water from the bottom of the reservoir. Bottom release is the normal mode of operation in these reservoirs, and the strength of stratification, the zone of withdrawal (zone of outflow), may be restricted to the hypolimnion, which results in a deeper epilimnion, warmer hypolimnion, and weaker density gradient.

Flood control reservoirs may be operated to maintain a constant pool elevation, or the pool elevation may be varied seasonally to provide additional flood storage. The timing and magnitude of the raising and drawdown can be adjusted within limits to manage, for example, in-lake fisheries. With respect to water quality, two points must be emphasized:

1. As the pool level increases, water flows back into the coves and embayments. The inflowing water and its associated nutrient load will probably be trapped in the cove until the pool level drops.
2. During periods of drawdown water trapped in dead-end coves is transported into the pelagic region of the lake. Pool level fluctuations may be the most efficient transport mechanism between these two regions of a reservoir.

It is also important to note that in actual operation the proposed reservoir elevation curve (rule curve) is only an objective, and the actual water surface elevation can vary significantly about the rule curve.

Many multipurpose reservoirs are equipped with multilevel outlet structures, making it possible to release water from different depths (Figure 2.13). It is therefore possible to meet temperature or other downstream water quality objectives by selectively withdrawing water from specific depths in the reservoir and blending it to achieve the desired objective. Selective withdrawal depends on stratification. If a reservoir is not stratified, classical potential flow theory is valid, and water flows radially from all depths toward the outlet. If the reservoir is stratified, the vertical buoyancy forces inhibit vertical motions, and the outflow zone is restricted to a horizontal layer that may be a few meters thick and extend the entire length of the reservoir. If water is being withdrawn from a well-mixed zone (i.e., the epilimnion), the withdrawal zone is restricted

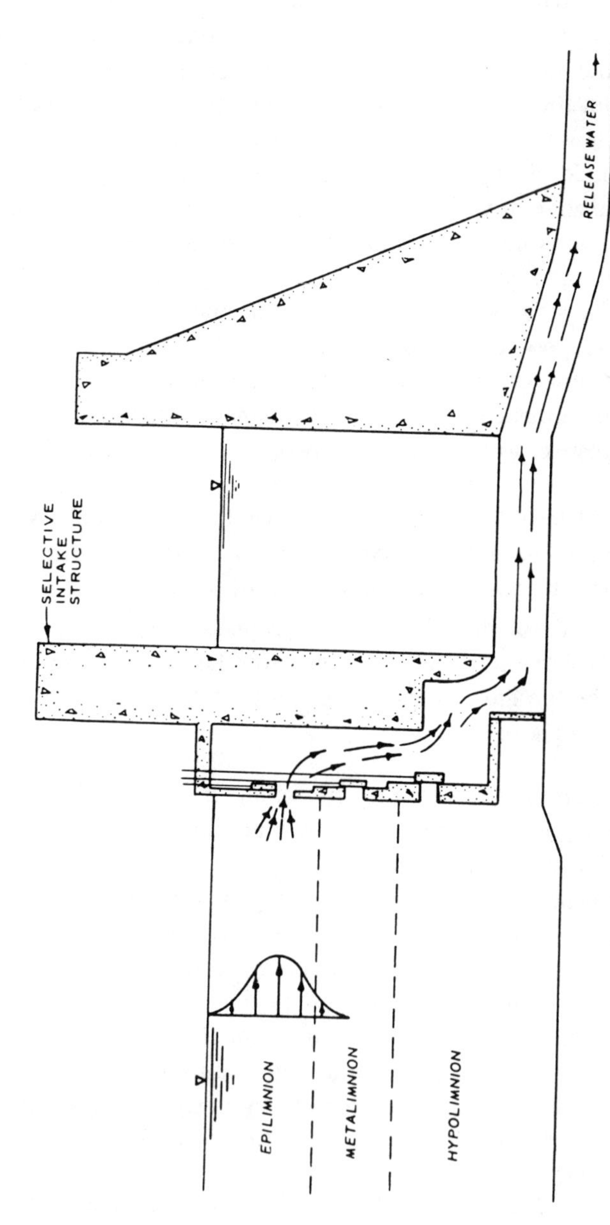

Figure 2.13 Example of a multilevel selective withdrawal outlet structure.

to the epilimnion until a critical flow is reached; then the vertical buoyancy forces are overcome and the withdrawal zone expands into the metalimnion. Complicated topography in the velocity of the outlet structure can modify the outflow zone from that predicted by simple models.

The outlet portals in a selective withdrawal structure are usually designed with a maximum flow capacity less than that required for flood flows. All of the flood flows are therefore released through the floodgates. These are sometimes located near the bottom of the reservoir. The actual operation is dependent on the rule curve, inflows, and downstream capacity.

Many reservoirs have minimum low-flow requirements. This flow must be released at all times and results in the pool elevation falling below the theoretical rule curve. During these periods the water level is dropping, and the overall movement of water and transport of constituents is out of the coves into the pelagic regions of the reservoir. Because the littoral zones may be a source of nutrients, this transport could impact the water quality of a reservoir. In addition, the minimum releases may increase mixing in the hypolimnion and promote the release of nutrients from anoxic sediments. In many instances the low flows are released from a special low-flow bypass located at a different elevation than the normal release level.

Because of their flexibility and capability to generate power instantly, hydropower projects are sometimes used for peaking power operations. Power is thus generated (releases are made) only during the hours of peak power demand. A typical peaking power schedule may be to release 600 m^3/s over a 4-hour period, Monday through Friday, and minimum releases only during the weekend. The impact of this type of operation on the mixing and resulting thermal structure is significant. For example, since the kinetic energy resulting from outflows is proportional to the flow rate to the third power (Q^3), the kinetic energy available for mixing from releasing 600 m^3/s over 4 hours (i.e., $KE \propto Q^3 \cdot \Delta t = (600)^3 \cdot 4$ hr $= 8.64 \times 10^8$) is 36 times larger than that from releasing 100 m^3/s over 24 hours (i.e., $KE \propto Q^3 \cdot \Delta t = (100)^3 \cdot 24$ hr $= 2.40 \times 10^7$), even though the total volume of water released is equal.

Pumped storage operations also increase mixing in hydropower reservoirs by pumping water back into the reservoir during off-peak hours of electrical power demand. A typical pumped storage project is shown in Figure 2.14. In this example power is generated from the upper, smaller reservoir to the main reservoir or to the downstream river. An example

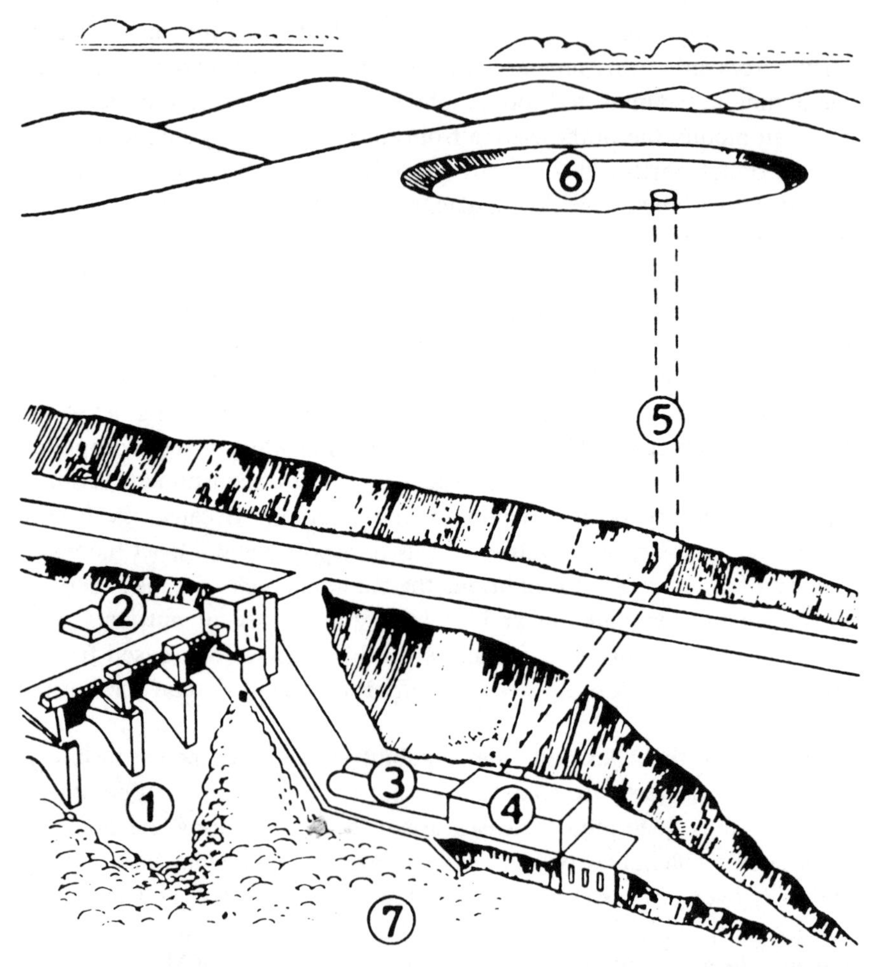

Figure 2.14 Schematic of Kinzua Dam hydroelectric facility. 1, Kinzua Dam; 2, Inlet–outlet structure, 3, main reservoir supply and return: Two 15-feet-diameter tunnels; 4, Seneca Power Plant; 5, supply and discharge tunnel: 22-feet diameter, $\frac{1}{2}$ mile long; 6, upper reservoir, 800 feet above plant; 7, Allegheny River.

of the impact of power generation on a reservoir thermal structure is shown in Figure 2.15. The higher hypolimnetic temperature reflects the increased level of hypolimnetic mixing with hydropower operations. This change in project operation also has a significant impact on dissolved oxygen.

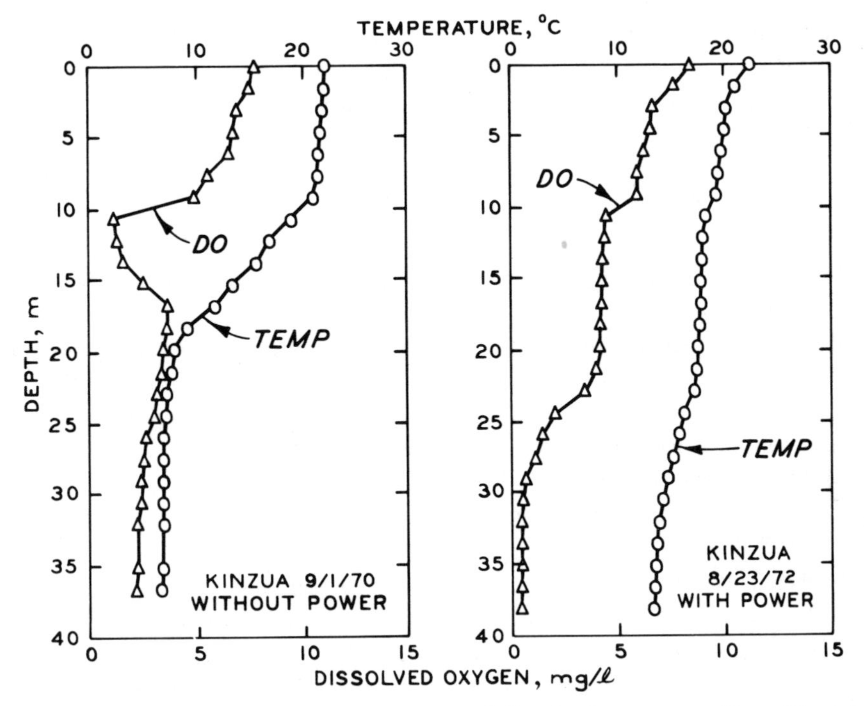

Figure 2.15 Comparison of temperature and dissolved oxygen profiles with and without hydropower generation.

Conclusions

The observed temperature structure in reservoirs is the synthesis of a number of complicated and interdependent phenomena. Stratification results from mixing and transport, but mixing and transport also depend on stratification.

Reservoirs are never in steady state. Meteorological forcing, inflows, outflows, and project operation are highly dynamic and characterized by a number of time scales.

Fluctuations and irregularities are just as important, or more important, than average quantities in determining transport and mixing in reservoirs. Turbulence and the resultant mixing are generated and reinforced by changes in project operation, wind, inflow, and complicated morphometry.

Waters of coves and embayments located off the main thalweg have a tendency to become isolated. The most efficient transport mechanism between these regions may be changes in water surface elevations.

All reservoirs are unique, and specific reservoirs differ from year to year as a result to hydrodynamics and transport.

REFERENCES

Bell, H. S. 1942. Density currents as agency for transporting fine sediments. *J. Geol.* 5:512–547.

Boyce, F. M. 1974. Some aspects of Great Lakes physics of importance to biological and chemical processes. *J. Fish. Res. Board Can.* 31:689–730.

Carmack, E. C. and C. R. J. Gary. 1982. Patterns of circulation and nutrient supply in a medium residence-time reservoir, Kootenay Lake, British Columbia. *Can. Wat. Res. J.* 7:51–69.

Csanady, G. T. 1975. Hydrodynamics of large lakes. *Ann. Rev. Fluid Mech.* 7:357–86.

Fischer, H. B., E. J. List, B. Y. C. Koh, J. Imberger, and N. H. Brooks. 1979. Mixing in inland and coastal waters. Academic Press, New York, NY.

Fischer, H. B. and P. D. Smith. 1983. Observation of transport to surface waters from a plunging inflow to Lake Mead. *Limnol. Oceangr.* 28:258–272.

Ford, D. E. and M. C. Johnson. 1981. Field observation of density currents in impoundments. Pages 1239–1248 in H. G. Stefan, ed. Proceedings of the symposium on surface water impoundments. Amer. Soc. Civil Engr., New York, NY.

Ford, D. E., M. C. Johnson, and S. G. Monismith. 1980. Density inflows to DeGray Lake, Arkansas. Pages 977–987 in T. Carstens and T. McClimans, eds. Proceedings of the second international symposium on stratified flows. Int. Assoc. for Hydraulic Res., Tapir, Trondheim, Norway.

Ford, D. E. and M. C. Johnson. 1983. Assessment of reservoir density currents and inflow processes. Tech. Rept. E-83-7. U.S. Army Engineer Waterways Experiment Station, CE, Vicksburg, MS.

Ford, D. E. and L. S. Johnson. 1986. An assessment of reservoir mixing processes. Tech. Report E-86-7. U.S. Army Engineer Waterways Experiment Station, CE, Vicksburg, MS.

Ford, D. E. and H. G. Stefan. 1980. Stratified variability in three morphometrically different lakes under identical meteorological forcing. *Wat. Res. Bull.* 16:243–247.

Grover, N. C. and C. S. Howard. 1938. The passage of turbid water through Lake Mead. *Trans. Am. Soc. Civ. Eng.* 103:720–732.

Hansen, N. O. 1978. Mixing processes in lakes. *Nordic Hydrology*. 9:57–74.

Hutchinson, G. E. 1957. A treatise on limnology. Vol. I Geography, physics and chemistry. John Wiley and Sons, New York, NY. 1015 pp.

Imberger, J. 1980. Selective withdrawal: A review. Pages 381–400 in T. Carstens and T. McClimans, eds. Proceedings of the second international symposium on stratified flow. Int. Assoc. for Hydraulic Res. Tapir, Trandheim, Norway.

Imberger, J. and P. F. Hamblin. 1982. Dynamics of lakes, reservoirs, and cooling ponds. *Ann. Rev. Fluid Mech.* 14:153–187.

Imberger, J. 1987. Mixing in river underflow. *Journal of Hydraulic Engineering,* 113(6). 697 pp.

Johnson, L. S. and Ford, D. E. 1987. Thermal modeling of DeGray Lake. Pages 536–564 in R. H. Kennedy and J. Nix, eds. Proceedings of the DeGray lake symposium Tech. Report E-87-4. U.S. Army Engineer Waterways Experiment Station, CE, Vicksburg, MS.

Johnson, T. R. 1987. Negatively buoyant flow in a diverging channel, Part 1: Flow Regimes. *Journal of Hydraulic Engineering* 113(6):716–730.

Knapp, R. T. 1942. Density currents: Their mixing characteristics and their effect on the turbulence structure of associated flow. Pages 289–306 in Proceedings of the second hydraulic conference, University of Iowa, Iowa City, IA.

Lindsey, R. K. and J. B. Franzini. 1972. Water resources engineering. McGraw-Hill Book Co. New York, NY.

Monismith, S., J. Imberger, and G. Billi. 1988. Shear waves and unsteady selective withdrawal. *Journal of Hydraulic Enginnering* 114(9). 1134 pp.

Mortimer, C. H. 1974. Lake hydrodynamics. *Mitt. Int. Ver. Limnol.* 20:124–97.

Smith, D. R., S. C. Wilhelms, J. P. Holland, M. S. Dortch, and J. E. Davis. 1987. Improved description of selective withdrawal through point sinks. Tech. Report E-87-2, U.S. Army Engineer Waterways Experiment Station, Vicksburg, MS.

Tennekes, H. and J. L. Lumley. 1972. A first course in turbulence. The MIT Press, Cambridge, MA.

Thornton, K. W., J. F. Nix, and J. D. Bragg. 1980. Coliforms and water quality: Use of data in project design and operation. *Wat. Res. Bull.* 16:86–92.

Wetzel, R. G. 1975. Limnology. W. B. Saunders Co., Philadelphia, PA. 767 pp.

CHAPTER 3

Sedimentary Processes

KENT W. THORNTON

Sediment transport and deposition is a dominant process in reservoirs that significantly influences the ecological response of the system. Considerations of the physical accumulation of sediment in reservoirs alone would indicate its potential importance in ecosystem structure and function. For reservoirs constructed prior to 1953 in the Midwest, the Great Plains states, and the southeastern and southwestern United States, 33% have lost from one-fourth to one-half of their original volume, about 14% have lost from one-half to three-quarters of their original volume, and about 10% have had all usable storage depleted by sediment deposition (Vanoni 1975). Sediment is not only the major water pollutant by weight and volume but is also a major carrier and catalyst for pesticides, organic residues, nutrients, and pathogenic organisms (Bachmann 1980, Ogg et. al. 1980, Sharpley et al. 1980, U.S. Senate Select Committee on Natural Water Resources 1960).

Watershed hydrology, sediment yield and transport, sediment discharge formulations and measurements, reservoir sediment surveys, and reservoir capacity computations are well documented by Chow (1964), Eagleson (1970), Thomas (1977), U.S. Geological Survey (1977), Vanoni (1975), Viessman et al. (1977) and others and will not be elaborated here. This chapter will focus instead on sedimentary processes that may result in a different response from reservoir ecosystems than in lake ecosystems. First, drainage basin characteristics and transport will be discussed; second, deposited and suspended sedimentation patterns and zones will be identified; and finally, the implications of these sedimentation patterns and processes in reservoirs will be discussed.

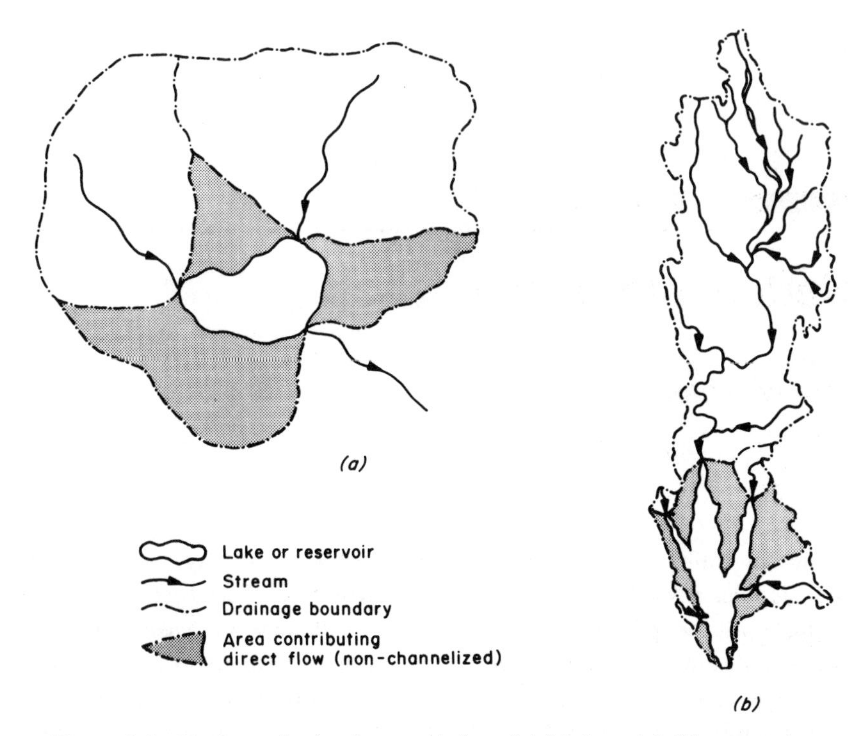

Figure 3.1 Drainage basin characteristics of (*a*) lakes and (*b*) reservoirs.

DRAINAGE BASIN CHARACTERISTICS AND SEDIMENT TRANSPORT PROCESSES

Since reservoirs are created by impounding lotic systems, the drainage basins for the reservoir and the upstream river are identical. In general, the stream order and size of the drainage basin upstream from reservoirs are higher and larger, respectively, than for lakes (Thornton et al. 1981). Reservoirs usually have the greatest proportion of the drainage basin located upstream from the impoundment. In contrast, many lakes have a considerable portion of contiguous drainage area (Figure 3.1). The shape and location of the drainage basin may influence the runoff and transport of material to lakes and reservoirs. Reservoir drainage basins are generally narrower and more elongated than lake watersheds, reflecting the riverine influence. Drainage basins generally become longer and narrower as they increase in size (Viessman et al. 1977).

While the process of watershed runoff is similar whether the receiving system is a lake or reservoir, differences in watershed characteristics between lakes and reservoirs influence the quantity and quality of material delivered to the system. Since there is a relation between the size of the drainage area and flow (Figure 3.2), such as:

$$Q = cA^{x}$$

where Q = flow as mean annual runoff
A = contributing drainage area
c, x = regression coefficients (Chow 1964),

larger drainage basins associated with reservoirs may result in greater annual flows entering reservoirs than lakes. Larger drainage basins and greater flows also indicate the potential for greater sediment and nutrient loads to reservoirs.

Watershed characteristics influence the sediment delivery ratio to the receiving system. There is a finite amount of energy in rainfall that determines the rate of erosion and transport of particulate matter from the watershed to the stream. As watershed size increases, the potential for interception and/or deposition of transported particulate matter increases, so the sediment delivery ratio is inversely proportional to the watershed area. However, since the relationship between drainage area and sediment delivery ratio is logarithmic and not linear, the absolute quantity of sediment and its adsorbed constituents delivered continues to increase with increased drainage area.

The finite energy of rainfall is also reflected in differential transport of particulate matter. Fluvial sediments are generally enriched in the finer clay and silt particles relative to the contributing watershed soils, since less energy is required to transport these finer particles from the watershed to the stream (Dendy 1981, Duffy et al. 1978, Rhoten et al. 1979). Additional sediment sorting and deposition occurs during stream transport (Vanoni 1975, Verhoff and Melfi 1978). Deposition of fluvial sediments in the floodplain is accompanied by the sorting of particulate material into coarse and fine particle sizes. The coarser, bed-load material generally accumulates in the stream channel while the finer material, silts and clays, accumulates on the floodplain (Vanoni 1975). This fine material, then, is readily washed into the stream during storm events and transported downstream to the reservoir. Fine silt and clay particles have a high sorptive capacity for phosphorus (Duffy et al. 1978, McCallister

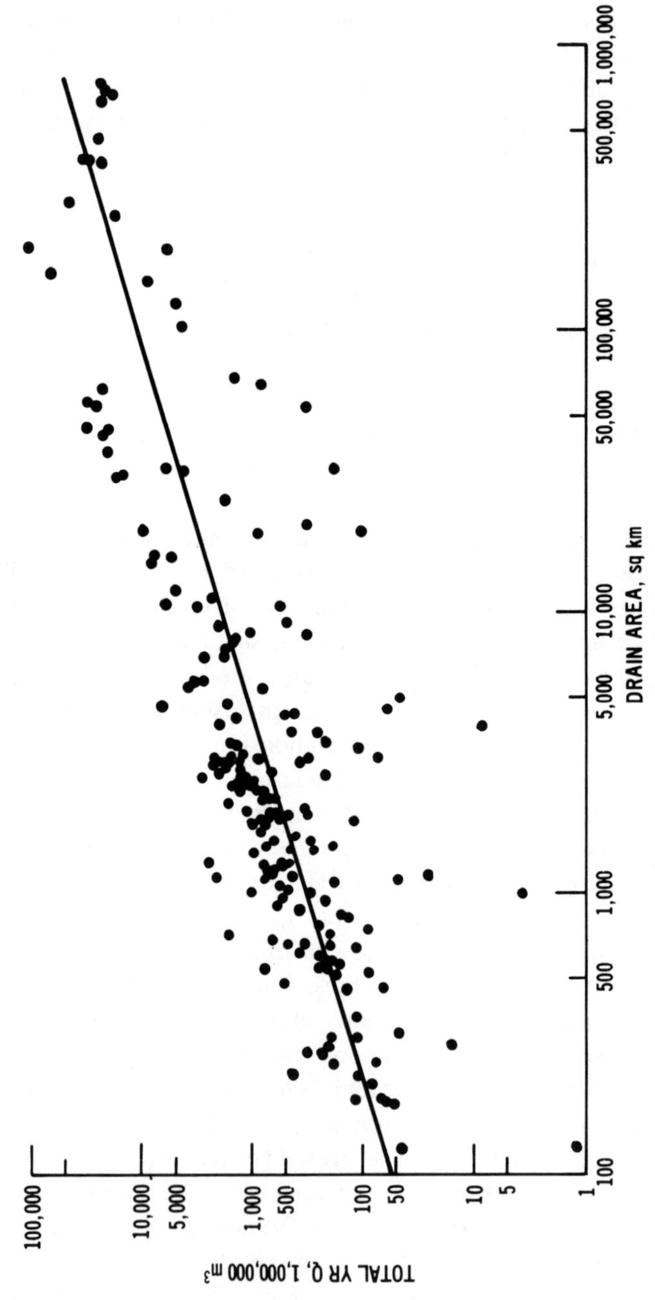

Figure 3.2 Increased total annual discharge as a function of increased reservoir drainage area (Data from Leidy and Jenkins 1977).

and Logan 1978, Schreiber and Rausch 1979, Sharpley and Seyers 1979, Sharpley et al. 1987), dissolved organic acids (Marzolf 1981), and other nutrients or contaminants (Pita and Hyne 1974). Because delivery to the stream is inversely proportional to the distance from the stream (Hebel 1981, Wilkin and Hebel 1982), land use in the floodplain has a greater impact on stream water quality than land use outside the flood plain. This leads to speculation that:

1. Total sediment delivery to reservoirs may be higher than to lakes.
2. The total mass of fine particles (e.g., silts and clays) is greater in the inflow to reservoirs than to natural lakes.
3. Corollary: The loading of particulate nutrients and other adsorbed material may be higher to reservoirs than to lakes.

The quality of material transported to reservoirs may also vary from natural lakes. Lakes are generally located in the upper portion of a drainage basin while reservoirs are generally located near the mouth of a drainage basin (Chapter 1, Figures 1.8*a* and 1.8*b*); thus, the stream order above lakes is generally lower than the stream order above reservoirs. Considerations of the river continuum concept (Minshall et al. 1985, Statzner and Higher 1985, Vannote et al. 1980) imply differences in the contribution of various forms of organic carbon to lakes and reservoirs. While there is a continuing discussion of the importance of heterotrophy vs. autotrophy in stream systems (Cummins 1974, Cummins et al. 1983, Fisher and Likens 1973, Minshall 1978, Minshall et al. 1985, Naiman and Sedell 1980, Statzner and Higler 1985, Vannottee et al. 1980), there is general agreement that autotrophic production tends to increase with stream order, at least up through a ninth-order stream (Naiman and Sedell 1981). Minshall (1978) has pointed out that autochthonous autotrophic production is an important contributor to the organic carbon supply in larger streams of forested areas and in open-canopy streams of all sizes.

Naiman and Sedell (1981) compared watersheds in Quebec and found that although 98 and 90% of the total number and total length, respectively, of streams were of first–third order, these streams represented only 20% of the stream surface area and yielded only 12% of the annual gross primary production of the entire drainage network. In contrast, streams of seventh–ninth order contributed 0.02 and 2.2% of the total number and total length, respectively, of the watershed streams but contributed 54% of the stream surface area and yielded 64% of the annual gross primary production of the entire drainage network (Naiman and Sedell 1981).

Sediment, particulate organic matter, and adsorbed constituents are transported primarily during storm events or elevated flows (Bilby and Likens 1979, Johnson et al. 1976, Kennedy et al. 1981, Sharpley and Seyers 1979, Verhoff and Melfi 1978). For rivers this transport may occur through a series of storm events with intermittent periods of deposition and processing in the stream between storm events (Verhoff and Melfi 1978, Verhoff et al. 1979). Since fine particulate organic matter (FPOM) is preferentially transported vs coarse particulate organic matter (POM) (Bilby and Likens 1979) and stream processing would be expected to increase the concentration of FPOM, reservoirs may also receive relatively higher proportions of FPOM and dissolved organic matter than natural lakes. The decrease in detrital standing stock with increasing stream order (Naiman and Sedell 1979), the decreasing input from leaf litter and processing with increasing stream order (Vannote et al. 1980), the increase in autotrophic production with stream order and watershed area, and the general distribution of reservoirs support this speculation of higher FPOM input to reservoirs

There also appears to be a strong seasonal component to suspended solids transport. This seasonality is a function of both watershed land-use and stream production. Tillage, planting, and other agricultural practices during the spring result in decreased interception of sediment and a lower percentage of clay particles in transported sediment. Schreiber et al. (1977) found clay content to be inversely related to sediment concentration from January to July. The clay component in sediment concentrations is generally highest from August to October, when the vegetation cover is maximal. This also corresponds to a period of increased organic phosphorus (P) content in suspended sediment concentrations (Schreiber and Rausch 1979) and the period when stream production may be maximal.

Finally, the spatial distribution of sediment loading is different between lakes and reservoirs. In lakes there is generally an equitable distribution of inflow around the periphery of the system (Figure 3.1). Reservoirs, in contrast, generally receive the majority of their inflow from one or two major tributaries located a considerable distance from the outflow. This promotes the development of pronounced physical and chemical gradients within reservoirs that have important consequences for reservoirs biological productivity and water quality.

SEDIMENTATION PATTERNS

Deltas

Water velocity and turbulence in the river begin to decrease in the headwater of the reservoir and result in decreased sediment-carrying capacity. High rates of particle deposition in the upper portion of reservoirs often result in the formation of a delta (Harrison 1983). Delta formation has generally been studied where rivers enter the sea or large lakes (Thankur and MacKay 1973). When rivers enter the broad, open expanses of an estuary or lake, the flow velocity, sediment load, and particle size distribution vary both laterally and longitudinally, generally resulting in a fan-shaped delta (Sundborg 1967). The longitudinal dimension is most important in reservoir delta formation. The river is generally confined to the old channel in the reservoir headwater so that the lateral dimension essentially remains constant and the velocity, sediment load, and particle size distribution vary longitudinally. The rate of delta progradation in lakes, then, is a two-dimensional problem with the longitudinal advance proportional to the square root of the delta area (Figure 3.3). In reservoirs the rate of delta progradation is a one-dimensional problem with the longitudinal advance directly proportional to the delta area (Figure 3.3).

Delta formation, migration, and reformation is a dynamic process that may change following each hydrometeorological event in both lakes and reservoirs. The rate of delta progradation, however, may be more rapid in reservoirs than in lakes. In Lake Mead, for example, the delta advanced 42 miles into the reservoir over a 13-year period from 1935 to 1948 (Vanoni 1975). Deposition of the coarse sediment in the delta and delta formation may influence sediment patterns not only in the reservoir but also in the upstream river.

Suspended Sediment Patterns

Suspended sediment and other particulate matter is transported primarily during storm events and elevated flows. However, few studies have been

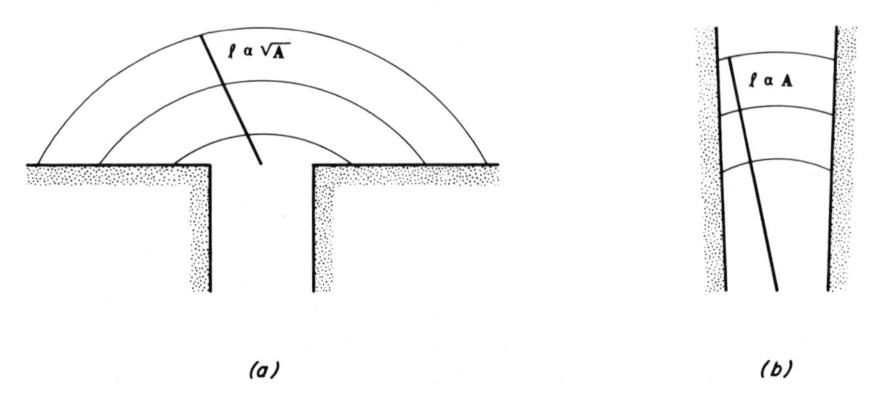

Figure 3.3 (*a*) Delta formation in lakes — a two-dimensional problem. (*b*) Delta formation in reservoirs — a one-dimensional problem.

conducted on the movement of storm flow and its associated load through reservoirs (Kennedy et al. 1981, Thornton et al. 1980). Low conductivity measurements associated with the inflow hydrograph provided an excellent tracer of storm flow through the reservoir Lake Red Rock (Kennedy et al. 1981). The storm flow exhibited significant short circuiting, moving along the old thalweg and passing through the reservoir in 2–3 days. The theoretical hydraulic residence times for this same period ranged from 5.5 to 9 days. Flow in the Des Moines River, the main tributary to Lake Red Rock, increased from a base flow of 141 m^3/s to 410 m^3/s. Inflowing mean suspended solids concentrations increased from 159 to 1702 mg/l. Suspended solids concentrations in the lower third of the reservoir, however, remained essentially unchanged throughout the storm indicating significant sedimentation in the upper end of the reservoir. The sediment trap efficiency, even with significant short circuiting, was still greater than 90%. Median particle diameters of suspended materials prior to the storm exhibited an exponential decrease from an inflow median diameter of ca. 10 μm to a median diameter of 3.5 μm near the dam. During the storm the headwater median particle size diameter decreased to ca. 6 μm and progressively increased to prestorm values as the storm flow passed. Progressive enrichment of fluvial suspended sediment by clay due to preferential erosion and transport of fine clays during runoff may explain the shift in particle size distribution in the river inflow (Dendy 1981, Rhoten et al. 1979). The median particle size diameter near the dam was relatively unaffected by the storm.

Storm flow in the reservoir DeGray Lake also followed the old thalweg, particularly when density flows proceeded as inter- and underflows (Thornton et al. 1980). Suspended solids concentrations were not measured, so sedimentation rates and patterns could not be determined explicitly. However, a two-dimensional hydrodynamic model used in combination with field turbidity data did provide additional insight into storm flow movement through DeGray Lake (Johnson et al. 1981). Model simulations indicated that sedimentation in the upper third of the reservoir was important in predicting observed field turbidities. Core samples of the deposited sediments later corroborated the importance of sedimentation in the upper third of DeGray Lake (Gunkel et al. 1984). The importance of storm events on reservoir and lake water quality is still relatively unknown.

Deposited Sediment Patterns

The longitudinal dimension is also important in sediment distributions within the reservoir. River inflow and its constituent sediment load generally follow the old river channel (thalweg); therefore, sediment deposition initially is greatest in the old channel. A sedimentation study conducted on Lake Red Rock, IA, seven years after impoundment revealed that approximately 8 m of sediment had been deposited in the channel in the upper portion of the reservoir with less than 1 m of deposition in the lower portion of the reservoir (Figure 3.4). Reservoirs exhibit plug-flow reactor characteristics with high sedimentation rates in the headwater, where sediment concentrations are highest, and an exponential decrease in sedimentation rate and sediment concentrations down the reservoir (Chapra 1981, Reckhow and Chapra 1983). There was an order of magnitude difference in sedimentation rates in Lake Red Rock, ranging from 19.1 cm/yr in the headwater to 1.4 cm/yr near the dam (Gunkel et al. 1984). McHenry et al. (1982) also found an exponential decrease in sedimentation rates in a long, narrow oxbow ranging from ca. 7 cm/yr near the inlet to less than 1 cm/yr near the outlet. Pharo and Carmack (1979) found similar reductions in sediment deposition rates in a riverine lake; i.e., 8 cm/yr near the delta to 0.34 cm/yr near the outflow.

Longitudinal gradients in sediment deposition also reflect the occurrence of longitudinal sorting of the particulate matter by particle size (McHenry et al. 1982, Olness and Rausch 1977, Pharo and Carmack 1979). The larger, heavier sands and coarse silts settle in the delta areas

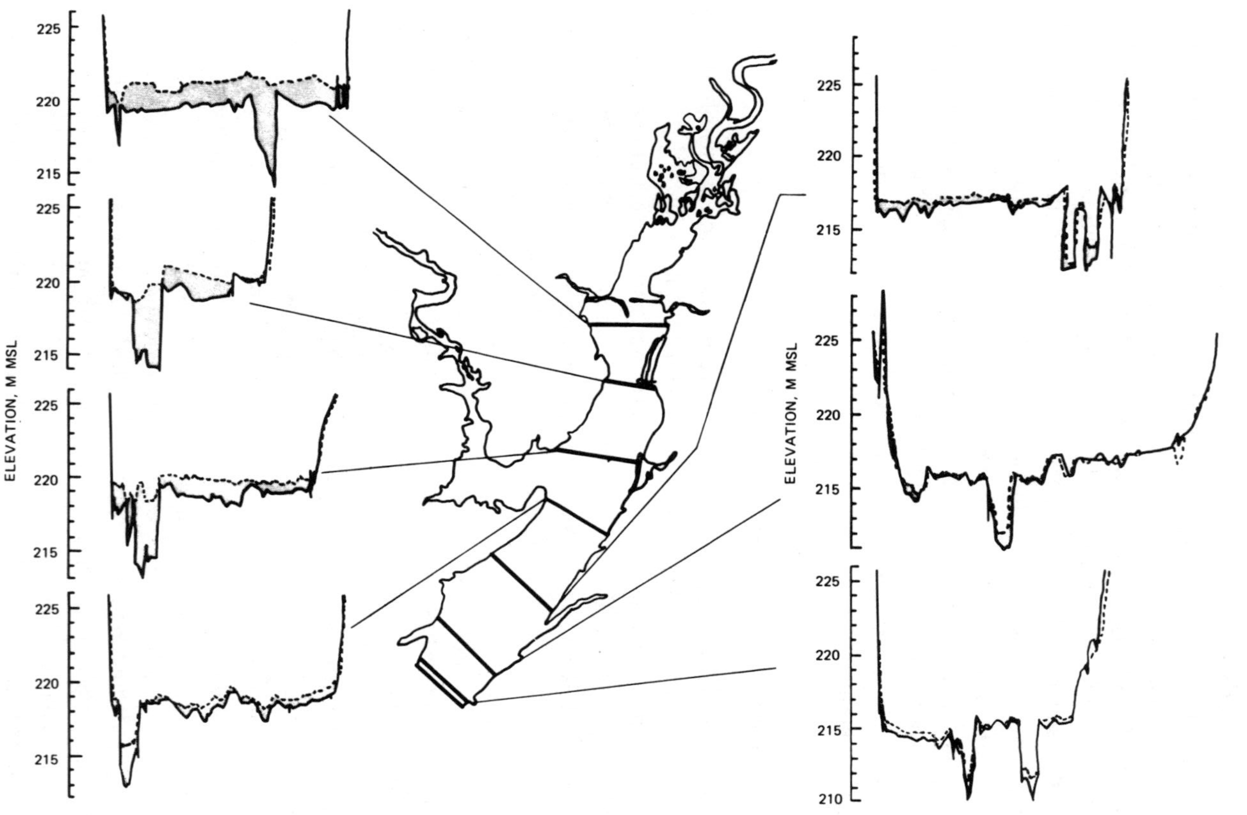

Figure 3.4 Sediment accumulation in the old channel and laterally across a transect from headwater (top of lake) to dam (bottom of lake) in Lake Red Rock, IA (from Gunkel et al. 1983).

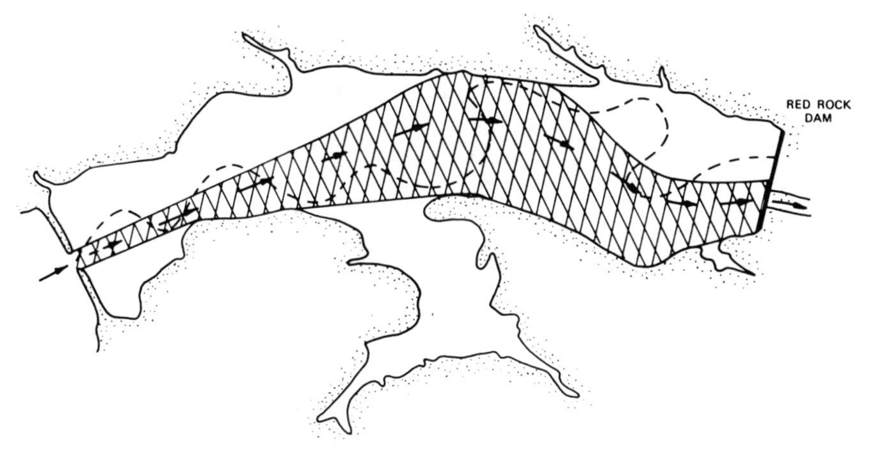

Figure 3.5 Generalized sedimentation rate distribution showing average flow patterns in a reservoir (after Gunkel et al. 1983).

of the reservoir (Figure 3.5). As flow velocities and turbulence continue to diminish, additional particle size sorting occurs along the longitudinal axis of the reservoir. The silts and coarse clays represent the next particle size class to settle with the fine clays and colloidal material, settling very slowly. The sediment particle size distribution in the Callahan Reservoir location varied from 5% sand, 76% silt, and 19% clay near the inlet to <1% sand, 61% silt, and 38% clay near midreservoir to 0% sand, 51% silt, and 49% clay near the outlet (Olness and Rausch 1977). A similar particle size distribution also occurred in Lake Red Rock sediments (Gunkel et al. 1984) (Figure 3.6).

The distributional pattern of deposited sediment may also provide indirect evidence of the 'average' flow conditions in the reservoir (Figure 3.5). Silts extend further down the reservoir over the old thalweg, reflecting the pattern of river flow through the reservoir (Kennedy et al. 1981). Deposited sediment patterns in a riverine lake indicated the importance of Coriolis effects on average flow pattern in the lake (Pharo and Carmack 1979).

Suspended-Deposited Sediment Interactions

Water control represents a major difference between lakes and reservoirs. Water control not only affects the zone and depth of withdrawal

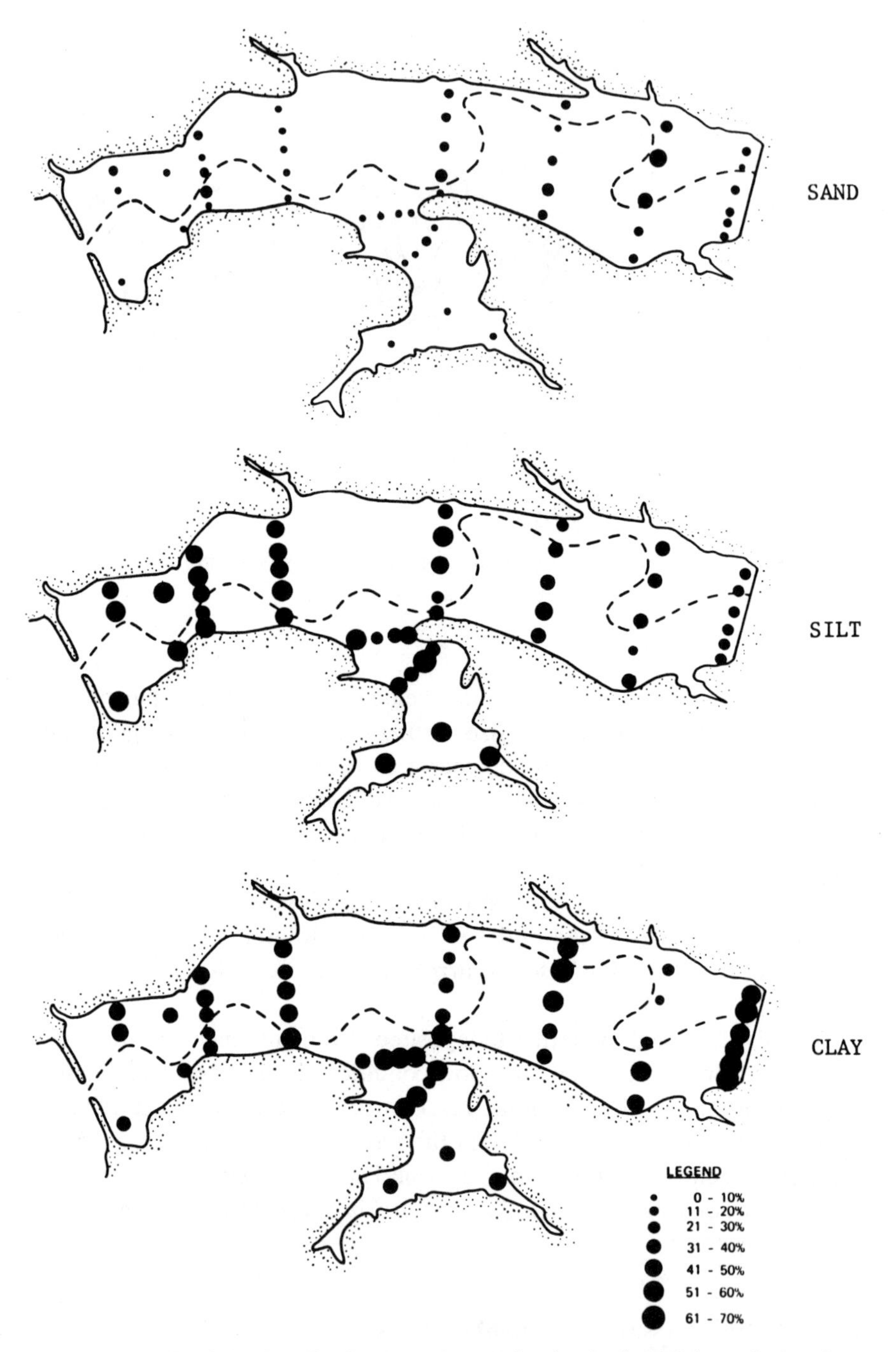

Figure 3.6 Particle size distribution of sand (top), silt (middle), and clay (bottom) in Lake Red Rock. Distribution of sand in the lower portion of the reservoir is due to sand bars deposited prior to impoundment (from Gunkel et al. 1983).

but also sediment distributions within reservoirs. Reservoir uses such as flood control, hydroelectric power generation, irrigation, and municipal and industrial water supply can result in significant water level fluctuations. These fluctuations may influence sedimentation patterns by altering reservoir morphometry (length, depth, volume, etc.), mixing regime, water exchange between coves and the main pool, water residence time, and other factors. During flood control operation and the storage of floodwaters Lake Red Rock can increase from 10 to 25 km in length, 10 to 22 m in depth, and 1.2 to $9.6 \times 10^8 m^3$ in volume. Big Eau Plaine Reservoir, WI, may fluctuate as much as 10 m in depth in conjunction with hydroelectric power generation (Kaster and Jacobi 1978). The rule curve for many reservoirs includes a drawdown period prior to elevated seasonal flows such as spring runoff.

DeGray Lake provides an example of the influence and interaction of morphometry and project operation on sedimentation in reservoirs. DeGray Lake is a peaking hydropower project in south-central Arkansas. Its selective withdrawal outlet is designed to permit the passing of flood flow through any of the withdrawal ports without having to operate the hypolimnetic floodgates. The elevation of the selective withdrawal outlet in DeGray Lake has generally been set in the epilimnion or near the top of the metalimnion. DeGray Lake is relatively narrow in the upper third of the reservoir and then rapidly expands in width (Figure 3.7). It has an average water residence time of 1.4 years with intermittent withdrawal for peaking power. The sudden expansion in width, long residence time (i.e., large storage), and intermittent meta- or epilimnetic withdrawal restricts suspended sediment deposition primarily to the upper half of the lake (Gunkel et al. 1984). Sedimentation has been so low in the lower half of the lake that sediment cores reveal the original soil and forest litter with only a thin overlayer of organic sediment. When no releases occur in DeGray Lake, interflows and their constituent loads stall in the upper portion of the reservoir around transect 14 (Figure 3.7) and disperse laterally. This is also the phenomenon that occurs in many lakes with no or little outflow. During storm flows or hydropower generation with large inflows or outflows, respectively, interflows and associated constituent loads may proceed over halfway through the reservoir before dispersing.

Water control, then, may significantly alter sedimentation patterns in a reservoir through redistribution, resuspension, and erosion of deposited sediment. Storage during elevated flow periods may result in delta formation and sedimentation occurring much farther upstream in the reservoir.

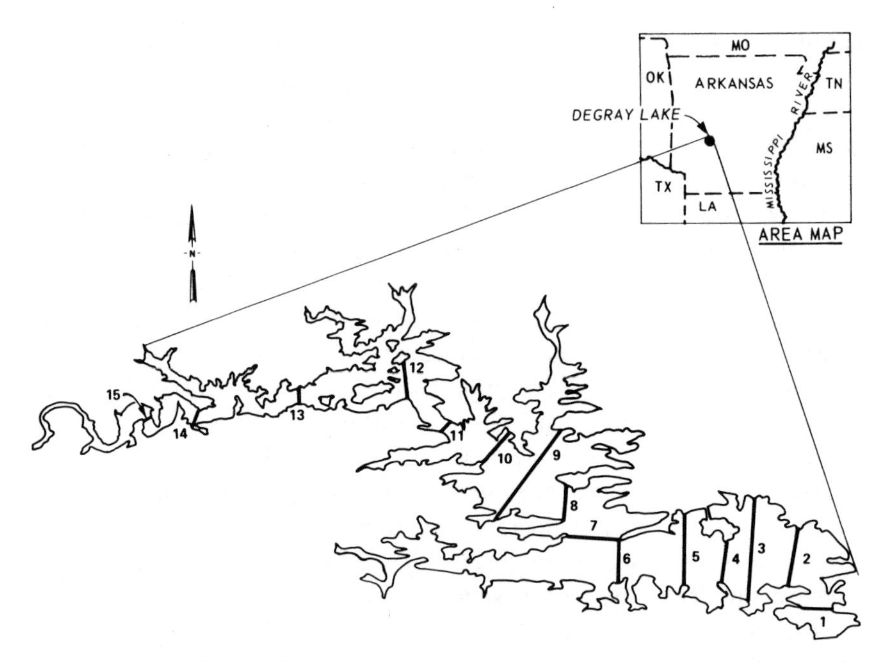

Figure 3.7 DeGray Lake showing the relatively narrow upper portion of the reservoir and the expansion below transect 11 (from Thornton et al. 1982).

As the pool returns to conservation or normal seasonal pool levels, sediment deposited, sorted, and processed upstream may be resuspended and transported downstream into the pool. As water recedes from the flood plain, finer particles may be resuspended and transported into the pool and redeposited. Similar particle size sorting may also occur with fluctuating pool elevations during withdrawal for water supply, irrigation, or hydropower operation.

As the water level falls, the finer particles deposited under a lower energy regime in the deeper, lower portion of the reservoir may become exposed to a higher energy regime. Direct and indirect turbulence from advective-and-wind-generated forces may resuspend deposited sediment not only from littoral areas but also from the pelagic zone. Bottom shear from density underflows may resuspend fine particles in the density current and transport them further into the reservoir. Fluctuating water levels may result in the development of extensive mudflat around the periphery of the reservoir. Continual exposure of the sediment to flooding and drying, wind and wave action, and weathering may also alter the particle size distribution within the sediment.

Resuspension influences not only the quantity of material redistributed in the reservoir but may also influence the water quality in the reservoir. It would, therefore, be beneficial to predict areas subject to deposition or accumulation and areas of resuspension or erosion. Hakanson (1977, 1981a, 1981b, 1982) has formulated a relationship for predicting resuspension based on lake area, mean depth, and maximum depth. The major variable in delineating areas of accumulation vs. erosion was the water content of the surficial sediments (Hakanson, Ibid.). These studies were related to simple basins with low shoreline development ratios, but it would be interesting to explore their applicability to reservoirs.

IMPLICATIONS

Longitudinal gradients in suspended sediment concentrations, sediment deposition, and particle size distributions clearly suggest the potential for chemical and biological gradients within the reservoir.

Delta and Riverine Zone

Coarse POM generally settles in the headwater region of the reservoir since flow velocities decrease rapidly in this riverine zone. Sediment samples from Whites Creek in the reservoir Watt Bar Lake, TN, contained leaves, stems, and twigs interlayed and mixed with clastic sedimentary layers (Worsley and Dennison 1973). Mandelbaum (1966) and Thornton et al. (1981) estimated that 60% of the inflowing sediment was retained in the St. Clair and Lake Red Rock deltas, respectively. This is also the zone where sedimentation of sand and coarse silt occurs, which should result in physical abrasion of coarse POM, much of which may be potamoplankton (i.e., plankton of rivers) or exported periphyton. Decreased velocities and turbulence in the river no longer maintain the algal cells in suspension, and high turbidity minimizes light penetration, so photosynthesis may be low in this zone (Cooper and Bacon 1981, Soballe 1981). These algae are generally thick-walled species or diatoms that can withstand abrasion during riverine transport but settle rapidly in a lower energy regime (Soballe 1981, Soballe and Kimmel 1987). In contrast to terrestrial detrital inputs to lakes from low-order streams, this suspended algal material represents a source of highly labile organic matter. Detrital inputs from terrestrial sources represent a more refractory source of organic matter, although microbial processing may result in a readily assimilative source of organic carbon. This continual source of labile organic matter should support a detritus-based foodweb of benthic shredders and grazers as well as an omnivorous fishery. Although community respiration may be high, the riverine zone is generally shallow, well-mixed, and therefore aerobic.

Upstream conditions may influence both the quantity and quality of organic matter and the development of the benthic community. Land use influences the loading of sediment and organic matter to streams, lakes, and reservoirs and has been extensively studied. Upstream impoundments and the disruption of the river continuum also influences this delivery (Ward and Stanford 1983, Soballe and Kimmel 1987). This is illustrated in two Oklahoma reservoirs receiving contrasting inputs of organic carbon. Lake Texoma has two major rivers, the Red and Washita rivers, rapidly prograding into the reservoir. These two rivers have essentially unregulated basins. Fort Gibson, however, is located on the Neosho River immediately downstream from two upstream impoundments in series. The organic carbon content in the headwater sediments of Fort Gibson is very low while the organic carbon content of headwater sediments of both river arms in Lake Texoma is relatively large (Hyne 1978). The disruption of the river continuum by reservoirs and their influence on downstream systems is presently an area of needed and active research.

Transition Zone

The transition zone is the area where silts, coarse-to-medium clays, and fine POM settle. While the sorptive capacity of these particles is not as high as that of the fine clays, there is still sorptive transport and settling of particulate phosphorus, organic carbon, iron (III), manganese (IV), calcium carbonate, and other elements. During stratification this zone has a relatively small hypolimnetic volume, and the biological processing of FPOM may rapidly deplete the hypolimnetic dissolved oxygen. Anaerobic conditions and redox reactions result in dentrification and resolubilization of adsorbed phosphorus, manganese (II), and iron (II) and sulfide release into the overlying water column (see Chapter 5). Anoxic conditions, then, may initially start in the upstream portion of the reservoir and proceed downstream throughout the period of stratification (Chapter 4).

The development of anoxic conditions in the transition zone has several potential implications for sedimentation and sediment-water interactions. Consider the factors listed below:

a) Flocculation of sediment particles increases with increasing conductivities or ionic strength in freshwater as well as in estuarine systems, for example, Lake Mead (Sherman 1953).
b) Stream sediments have much greater sorption maxima than the associated watershed soils. McCallister and Logan (1978) attributed this characteristic not only to a preferential enrichment of clay but also to chemical alteration of iron-substituted clays after exposure to anoxic conditions in the stream had been followed by resuspension, oxidation, and hydration of the iron-clay particles.
c) Clay particles sorb not only metals and nutrients but also dissolved organic compounds (DePinto et al. 1982, Frink 1969, Pita and Hyne 1974).

This leads to the following scenario for processes occurring in the transition zone.

As anoxic conditions develop at the head of the transition zone, decreased redox potential results in an increase in ammonium–N concentrations and the resolubilization of manganese, iron, and phosphorus adsorbed to particulate matter. This increases the ionic strength and concentration of dissolved constituents in the water column and results in increased flocculation and sedimentation of finer sediment particles. As this material settles through the anoxic zone, adsorbed constituents are reduced, further increasing the ionic strength and dissolved constituent concentration of the water column. A feed-forward loop is, therefore, initiated in this zone as increased sedimentation → increased organic matter for decomposition → increased dissolved constituent concentrations → increased ionic strength → increased sedimentation.

One of the characteristics of the transition zone is its dynamic nature. Since it is strongly influenced by river inflow and reservoir discharge, the anoxic zone may move several kilometers into the reservoir, several kilometers back upstream, or be completely dissipated in a day. Based on the concentration gradient of dissolved constituents between the anoxic hypolimnion and meta-epilimnion, molecular diffusion alone could result in a significant flux of phosphorus, manganese, and iron out of the anoxic zone. Considering the influence of wind and advective forces, turbulent diffusion may greatly increase the fluxes of both dissolved and particulate materials. As these clay-metal complexes are transported out of the anoxic zone, oxidized and hydrated, their sorptive capacities may significantly increase. One (1) gram of freshly oxidized MnO_4, for example, has a surface area of 300 m^2 (Nix, personal communication). Manganese, with a +4 valence and a small colloidal particle size, may actually be more effective at scavenging nutrients and other anions from the water column than the larger iron (III) particles, which settle rapidly. Manganese may be advectively transported considerable distances in a reservoir, even during low-flow periods (Nix 1981a). Sedimentation of the iron-manganese-clay complexes is reduced by turbulence, so these particles may be retained in the water column for a considerable time. This longer exposure in the water column may also increase the quantity of material sorbed on particle surfaces.

Finally, as these finer particles are transported and oxidized, they may immediately flocculate to form larger particles that settle just downstream from the leading edge of the anoxic zone. Sherman (1953) found that particles larger than 24 μm generally settled as individual particles while particles smaller than 24 μm formed flocculated aggregates. Immediate sedimentation of these aggregated particles, their adsorbed constituents, and the associated microbial assemblage could contribute to the hypolimnetic oxygen demand and continue to move the anoxic zone downstream. Laboratory and field experimentation should determine whether this speculation has substance.

Lacustrine Zone

The lacustrine zone in the lower portion of the reservoir has sedimentation patterns reflecting the settling of both fine clays and colloidal material as well as the autochthonous (within-reservoir) production of organic matter. In both Fort Gibson and Lake Texoma high organic carbon values were concentrated along the channel in the deeper areas near the dam (Hyne 1978). Whether the oxygen demand exerted by this particulate matter contributes to an anoxic zone near the dam is a function of reservoir morphometry, hydrometeorology, and operation (see Chapter 4). Reservoirs with relatively little fetch, moderate depth (~20 m), moderate to strong thermal stratification, and operated for warm-water releases [such as three southwest Arkansas reservoirs (Nix 1981b)] may develop an anoxic zone near the dam. This anoxic zone may expand both vertically up through the water column and upstream and may combine with the anoxic zone moving downstream from the transition zone (Nix 1981b). The particulate matter oxygen demand may be satisfied in passage through the water column in deep reservoirs. Shallow reservoirs or

Table 3.1 Reservoirs and riverine lakes exhibiting longitudinal gradients in particulate size distribution and chemical composition (i.e., organic C, phosphorus, metals, etc.)

Reservoir	Element	Investigator
Callahan Reservoir	P	Olness and Rausch (1977)
DeGray Lake	P, Fe, Mn	Gunkel et al. (1984)
		Thornton et al. (1982)
Eau Galle Lake	P, Fe, Mn	Gunkel et al. (1984)
Kamloops Lake	Fe	Pharo and Carmack (1979)
Lake Red Rock	Fe, Mn	Thornton et al. (1981)
Lake Texoma	organic C	Hyne (1979)
Watts Bar Lake	organic detritus	Worsley and Dennison (1973)
West Point Lake	P, Fe, Mn	Gunkel et al. (1984)
Wolf Lake	P	McHenry et al. (1982)

moderately deep reservoirs with bottom withdrawal (i.e., Canyon Reservoir, see Chapter 4) that have relatively large fetches and weak to moderate thermal stratification may have sufficient hypolimnetic mixing from seiches, and advective and/or wind-generated forces to minimize anoxic development near the dam.

The development of anoxic conditions may result in the release of significant concentrations of iron, manganese, phosphorus, and other elements adsorbed to clay particles. An inverse relation between particle size and the percent of acid-extractable iron was reported for sediments of a riverine lake, where the highest percentages of iron and smallest median particle size diameters were found near the outlet (Pharo and Carmack 1978). Similar sediment and elemental patterns are found commonly in reservoirs and riverine lakes (Table 3.1).

Sedimentation may also contribute to the development of a metalimnetic oxygen minimum. Nix (personal communication) has found that the magnitude of the metalimnetic minimum in DeGrey Lake depends on the sequence of elevated spring flows and the onset of thermal stratification. If the spring flows precede the development of stratification, the organic loadings associated with these flows are generally dispersed throughout the water column. If stratification has begun before the elevated spring flows, organic loadings may enter the metalimnion via interflow and result in greater metalimnetic oxygen demand.

The importance of organic loadings to reservoir metabolism needs to be investigated as a function of drainage area characteristics, hydrometeorology, reservoir morphometry, and water control. Lind (1971) determined that autochthonous production in Waco Reservoir was about three times the allochthonous contribution to the annual energy budget. In Lake Red Rock, however, suspended particulate matter — phytoplankton interactions resulted in equivalent production and respiration (P:R = 1.3) in the riverine zone but a heterotrophic environment (P:R = 0.4) in the downlake portion of the reservoir. Allochthonous organic contributions in both Kuybyshev and Rybinsk Reservoirs were about three times greater than autochthonous contributions, regardless of the level of production of organic matter in the reservoir (Ivatin 1974).

Fluctuating Water Level

Water level fluctuations, including winter drawdown, represent a major abiotic stress to reservoir ecosystems (Ryder 1978). The flooding and dewatering of soils and associated changes in sediment composition and chemistry have received considerable study by agronomists and other scientists working on economic crops such as rice, and Whitlow and Harris (1979) have applied this information to reservoirs. One of the major impacts of flooding is the rapid conversion from aerobic to anaerobic sediments (i.e., less than one day, Whitlow and Harris 1979). While the importance of migration of resolubilized elements such as ammonium −N, phosphorus, iron, and manganese to the sediment surface is uncertain, sediment, erosion or resuspension by wind or wave action may expose the overlying water to relatively high concentrations of these elements. High winds, associated with the passage of weather fronts, resuspended deposited sediment to as much as 1 m in depth in Lake Carl Blackwell, Oklahoma (Norton 1968). Alternating periods of flooding, dewatering, and resuspension may result in significant transport, exchange, and redeposition of sediments and their associated constituents in reservoirs and enhanced interactions with the overlying water column (Baxter 1985).

Nutrient Loading Models

An important coefficient in many nutrient loading models is the sedimentation rate or settling velocity coefficient. Recent nutrient loading models have stressed the difference between sedimentation patterns in lakes and reservoirs (Chapra 1981, Higgins and Kim 1981, Mueller 1982). Most lake models assume the lake is a completely mixed reactor while recent reservoir models assume the reservoir to be a plug-flow reactor. Neither assumption is generally met in actual application; however, theoretical differences between these two models are important in comparing sedimentation patterns in lakes and reservoirs. Reservoirs exhibit plug-flow reactor characteristics with high sedimentation rates in the headwater, where suspended sediment concentrations are highest, and an exponential decrease in sedimentation down the reservoir axis as suspended sediment concentrations also decrease. These longitudinal gradients in sedimentation also result in a longitudinal gradient in reservoir or riverine lake trophic state from eutrophic in the headwater to oligotrophic at the dam (Hannan et al. 1981, Peters 1979, Thornton et al. 1981, 1982, Kimmel and Groeger 1984, Chapter 4). Chapra (1981) has presented an example of considering solids in the dynamics of reservoirs of river-run lakes where removal of suspended solids may significantly decrease the water quality of the system. The direct application of many nutrient loading models developed from lake databases to reservoirs without testing or modification may not be warranted.

SPECULATION

The primary purpose of this discussion is to indicate the importance of sediment-water interactions on both reservoir processes and ecosystem responses. Although sedimentary processes obviously play an important role in reservoir limnology, we still know relatively little about these processes in reservoirs. Since the purpose of this book is to stimulate discussion of limnological processes in reservoirs, we offer the following summary statements for consideration:

1. Landuse on the floodplain dictates the loading of sediments and adsorbed constituents to reservoirs. The impact of land use outside the floodplain decreases as a power function of the distance from the stream.
2. There is a seasonal component to the loading of clays and organic matter to reservoirs that significantly influences the response of the

system. Allochthonous organic loadings may represent an important source of carbon to reservoirs.

3. Differential sedimentation from headwater to dam results not only in longitudinal gradients in particle sizes but also in longitudinal gradients in chemical constituents and biotic assemblages.
4. The development of anoxic conditions in the transition zone may set up a feed-forward loop that promotes additional sedimentation and chemical alteration of particles.
5. Sorption-desorption reactions of suspended clay particles may remove nutrients, organic carbon, and metals from the mixed layer.
6. Fluctuating water levels may significantly enhance sediment-water interactions in reservoirs and result in increased solids and nutrient concentrations in the water column.
7. Fluctuating water levels promote the exchange of sediments and adsorbed constituents between coves and the main pool.
8. Plug-flow nutrient loading models should describe reservoir trophic characteristics better than models developed from lake databases.

REFERENCES

Bachmann, R. W. 1980. The role of agricultural sediments and chemicals in eutrophication. *Wat. Pollut. Contr. Fed.* 52:2425–2437.

Bilby, R. E. and G. E. Likens. 1979. Effect of hydrologic fluctuations on the transport of fine particulate organic carbon in a small stream. *Limnol. Oceanogr.* 24:69–75.

Chapra, S. C. 1981. Application of phosphorus loading models to river-run lakes and other incompletely mixed systems. Pages 329–334 in Restoration of lakes and inland waters. EPA 440/5-281-010, United States Environmental Protection Agnecy.

Chow, V. T., ed. 1964. Handbook of applied hydrology. McGraw-Hill Book C. New York, NY.

Cooper, C. M. and E. J. Bacon. 1981. Effects of suspended solids on primary productivity in Lake Chicot, Arkansas. Pages 1357–1367 in H. G. Stefan, ed. Proceedings of the symposium on surface water impoundments. Amer. Soc. Civil Engr. New York, NY.

Cummins, K. W. 1974. Structure and function of stream ecosystems. *Bioscience.* 64:631–641.

Cummins, K. W., J. R. Sedell, F. J. Swanson, G. W. Minshall, S. G. Fisher, C. E. Cushing, R. C. Petersen, and R. L. Vannote. 1983. Organic matter budgets for stream ecosystems: Problems in their evaluation, pages 299–353. In J. R. Barnes and G. W. Minshall eds. Stream ecology: Application and testing of general ecological theory. Plenum Press, New York, NY. 399 p.

Dendy, F. E. 1981. Sediment yield from a Mississippi delta cotton field. *Env. Qual.* 10:482.

DePinto, J. V., T. C. Young, and S. C. Martin. 1982. Aquatic sediments. *Wat. Pollut. Contr. Fed.* 54:855–862.

Duffy, P. D, J. D. Schreiber, D. C. McClurkin, and L. L. McDowell. 1978. Aqueous and sediment — phase phosphorus yields from five southern pine watersheds. *Env. Qual.* 7:45–50.

Eagleson, P. S. 1970. Dynamic hydrology. McGraw-Hill Book Co., New York, NY.

Fisher, S. G. and G. E. Likens. 1973. Energy flow in Bear Brook, New Hampshire: An integrative approach to stream ecosystem metabolism. *Ecol. Monogr.* 43:421–439.

Frink, C. R. 1969. Chemical and mineralogical characteristics of eutrophic lake sediments. *Soil Sci. Soc. Am. Proc.* 33:369–372.

Gunkel, R. C., R. F. Gaugush, R. H. Kennedy, G. E. Sand, J. H. Carroll, and J. Cauthey. 1984. A comparative study of sediment quality in four reservoirs. Technical Report E-84-2. U.S. Army Engineers Waterways Experiment Station, Vicksburg, MS.

Hakanson, L. 1977. The influence of wind, fetch, and water depth on the distribution of sediments in Lake Vanern, Sweden. *Can. Earth Sci.* 14:379–412.

Hakanson, L. 1981a. On lake bottom dynamics — the energy — topography factor. *Can. Earth Sci.* 18:899–909.

Hakanson, L. 1981b. Determination of characteristic values for physical and chemical lake sediment parameters. *Wat. Resour. Res.* 17:1625–1640.

Hakanson, L. 1982. Lake bottom dynamics and morphometry — the dynamic ratio. *Wat. Resour. Res.* 18. 1444–1450.

Hannan, H. H., D. Burrows, and D. C. Whitenberg. 1981. The trophic status of a deep-storage reservoir in central Texas. Pages 425–434. In H. G. Stefan, ed. Proceedings of the symposium on surface water impoundments. *Amer. Soc. Civil Engr.* New York.

Hebel, S. J. 1981. The use of fallout Cesium-137 to determine patterns of soil movement and their implications for land management and water quality planning. M.S. Thesis., University of Illinois, Urbana, IL.

Higgins, J. M. and B. R. Kim. 1981. Phosphorus retention models for Tennessee Valley Authority reservoirs. *Water Resour. Res.* 17:571–576.

Hyne, N. J. 1978. The distribution and source of organic matter in reservoir sediments. *Env. Geol.* 2:279–287.

Ivatin, A. V. 1974. Production of phytoplankton and decomposition of organic matter in the Kuybyshev Reservoir. *Hydrobiol.* 10:49–52.

Johnson, A. H., D. R. Bouldin, E. A. Joyette, and A. M. Hodges. 1976. Phosphorus loss by stream transport from a rural watershed: Quantities, processes, and sources. *Env. Qual.* 5:148–157.

Johnson, M. C., D. E. Ford, E. M. Buchak, and J. E. Edinger. 1981. Analyzing storm event data from DeGray Lake, Arkansas using LARM. Amer. Soc. Civil Engr. Presentation, Fall Meeting. St. Louis, MO.

Kaster, J. L. and G. Z. Jacobi. 1978. Benthic macroinvertebrates of a fluctuating reservoir. *Freshwat. Biol.* 8:283–290.

Kennedy, R. H., K. W. Thornton, and J. H. Carroll. 1981. Suspended sediment gradients in Lake Red Rock. Pages 1318–1328 in H. G. Stefan, ed. Proceedings of the symposium on surface water impoundments. Amer. Soc. Civil Engr. New York, NY.

Kimmel, B. L. and A. W. Groeger. 1984. Organic matter supply and processing in lakes and reservoirs. NALMS Proceedings in Lake and Reser. Mgt. U.S. EPA 440/5/84-001. p. 277–281.

Lind, O. T. 1971. The organic matter budget of a central Texas reservoir. Pages 193–202 in G. E. Hall, ed. Reservoir fisheries and limnology. Am. Fish. Soc.

Mandelbaum, H. 1966. Sedimentation in the St. Clair River delta. *Great Lakes Res. Div. Pub.* 15:192–202.

Marzolf, G. R. 1981. Some aspects of zooplankton existence in surface water impoundments. Pages 1392–1399 in H. G. Stefan, ed. Proceedings of the symposium on surface water impoundments. Amer. Soc. Civil Engr. New York, NY.

McCallister, D. L. and T. J. Logan. 1978. Phosphate adsorption-desorption characteristics of soils and bottom sediments in the Maumee River basin of Ohio. *Env. Qual.* 7:87–92.

McHenry, J. R., C. M. Cooper, and J. C. Ritchie. 1982. Sedimentation in Wolf Lake, lower Yazoo river basin, Mississippi. *Freshwat. Ecol.* 1:547–558.

Minshall, G. W. 1978. Autotrophy in stream ecosystems. Bioscience 28:767–771.

Minshall, G. W., K. W. Cummins, R. C. Peterson, C. E. Cushing, D. A. Bruns, J. R. Sedell, and R. L. Vannote. 1985. Developments in stream ecosystem theory. *Can. J. Fish. Aquat. Sci.* 42:1045–1055.

Mueller, D. K. 1982. Mass balance model estimation of phosphorus concentrations in reservoirs. *Wat. Resour. Bull.* 18:377–382.

Naiman, R. J. and J. R. Sedell. 1981. Stream ecosystem research in watershed perspective. *Verb. Int. Verein. Limnol.* 21:804–811.

Naiman, R. J. 1980. Relationships between metabolic parameters and stream order in Oregon. *Can. Jour. Fish Aquat. Sci.* 37:834–847.

Naiman, R. J. and J. R. Sedell. 1979. Characterization of particulate organic matter transported by some Cascade Mountain streams. JFRBC 36: 17–31.

Nix, J. 1981a. Contribution of hypolimnetic water on metalimnetic dissolved oxygen minima in a reservoir. *Wat. Resour. Res.* 17:329–332.

Nix, J. L. 1981b. Report on water quality reconnaissance of Lakes Dierks, DeQueen, Gillham, and Millwood, Arkansas. U.S. Army Engineer Little Rock District. 34 pp.

Norton, J. L. 1968. Distribution character and abundance of sediment in a 3000 acre impoundment in Payne County, Oklahoma. M. S. Thesis, Oklahoma State University, Stillwater, OK.

Ogg, C. W., R. Heimlich, and H. Pionke. 1980. Efficiently reducing nonpoint phosphorus loads to lakes and reservoirs. *Wat. Resour. Bull.* 16:967–970.

Olness, A. and D. Rausch. 1977. Callahan Reservoir: III. Bottom sediment-water-phosphorus relationships. *Trans. Am. Soc. Ag. Engr.* 20:291–297, 300.

Peters, R. H. 1979. Concentration and kinetics of phosphorus fractions along the trophic gradients of Lake Memphremagog. *J. Fish. Res. Board. Con.* 36:970–979.

Pharo, C. H. and E. C. Carmack. 1979. Sedimentation processes in a short residence-time intermontane lake, Kamlooops Lake, British Columbia. *Sedimentology*. 26:523–541.

Pita, F. W. and N. J. Hyne. 1974. The depositional environment of zinc, lead and cadmium in reservoir sediments. *Wat. Res.* 9:701–706.

Reckhow, K. H. and S. C. Chapra. 1983. Engineering approaches for lake mangement. Data analysis and empirical modeling. Ann Arbor Science, Ann Arbor, MI. 340 pp.

Rhoten, F. E., N. E. Smeck, and L. P. Wilding. 1979. Preferential clay mineral erosion from watersheds in the Maumee River basin. *Env. Qual.* 8:547–550.

Ryder, R. A. 1978. Ecological heterogeneity between north-temperate reservoirs and glacial lake systems due to differing succession rates and cultural uses. *Verh. Int. Verein. Limnol.* 20:1568–1574.

Schreiber, J. D. and D. L. Rausch. 1979. Suspended sediment-phosphorus relationships for the inflow and outflow of a flood detention reservoir. *Jour. Env. Qual.* 8:510–514.

Schreiber, J. D., D. L. Rausch, and L. L. McDowell. 1977. Callahan Reservoir. II. Inflow and outflow suspended sediment phosphorus relationships. *Trans. Am. Soc. Ag. Engr.* 20:285–290.

Sharpley, A. N., S. J. Smith, R. G. Menzel, W. A. Berg, and O. R. Jones. 1987. Precipitation and water quality in the Southern Plains. NALMS Proceedings in Lake and Reser. Mgt. Vol. 3, 379–384.

Sharpley, A. N. and J. K. Syers. 1979. Phosphorus inputs into a stream draining an agricultural watered. II: Amounts contributed and relative significance of runoff types. *Wat. Air. Soil Pollut.* 11:417–428.

Sherman, I. 1953. Flocculent structure of sediment suspended in Lake Mead. *Trans. Am. Geophys. Union.* 34:394–406.

Soballe, D. M. 1981. The fate of river phytoplankton in Red Rock Reservoir. Ph.D. Thesis, Iowa State University, Ames, IA.

Soballe, D. M. and B. L. Kimmel. 1987. A large scale comparison of factors influencing phytoplankton abundance in rivers, lakes, and impoundments. *Ecology* 68:1943–1954.

Sundborg, A. 1967. Some aspects of fluvial sediments and fluvial morphology 1. General views and graphic methods. *Geografiska Annal.* 49A:333–343.

Statzner, B. and B. Higler. 1985. Questions and comments on the River Continuum Concept. *Can. J. Fish. Aquat. Sci.* 42:1038–1044.

Thakur, T. R. and D. K. MacKay. 1973. Delta processes. Pages 509–530 in Proceedings of the Hydrol. symp., fluvial processes and sedimentation. Nat. Res. Council, Ottawa, Canada.

Thomas, W. A. 1977. Sediment transport, HEC-IHD-1200, Vol. 12. U.S. Army Engineer Hydrologic Engineering Center, Davis, CA.

Thornton, K. W., R. H. Kennedy, J. H. Carroll, W. W. Walker, R. C. Gunkel, and S. Ashby. 1981. Reservoir sedimentation and water quality — an heuristic model. Pages 654–661 in H. G. Stefan, ed. Proceedings of the symposium on surface water impoundments. *Amer. Soc. Civil Engr.* New York, NY.

Thornton, K. W., J. F. Nix, and J. D. Bragg. 1980. Coliforms and water quality: Use of data in project design and operation. *Wat. Resour. Bull.* 16:86–92.

U.S. Geological Survey. 1977. National handbook of recommended methods for water-data acquisition. Office of Water Data Coordination. U.S. Department of the Interior, Reston, VA.

U.S. Senate Select Committee on National Water Resources. 1960. Pollution Abatement Committee Print No. 9. 86th Congress. 2nd Session.

Vannote, R. L., G. W. Minshall, K. W. Cummins, J. R. Sedell, and C. E. Cushing. 1980. The river continuum concept. *Can. Jour. Fish. Aquat. Sci.* 37:130–137.

Vanoni, V. A., ed. 1975. Sedimentation engineering. Amer. Soc. Civil Engr. New York, NY.

Viessman, W., Jr., J. W. Knapp. G. L. Lewis, and T. E. Harbaugh. 1977. Introduction to hydrology. Harper and Row Publishers, New York, NY.

Verhoff, F. H., D. A. Melfi, and S. M. Yaksich. 1979. Storm travel distance calculations for total phosphorus and suspended materials in rivers. *Wat. Resour. Res.* 15:1354–1360.

Verhoff, F. H. and D. A. Melfi. 1978. Total phosphorus transport during storm events. *J. Env. Engr. Amer. Soc. Civil Engr.* 104:1021–1023.

Ward J. W. and J. A. Stanford. 1983. The serial discontinuity concept of lotic ecosystems, pages 29–42. In T. D. Fontaine and S. M. Bartell, eds. Dynamics of lotic ecosystems. Ann Arbor Science Publishers Inc., Ann Arbor, MI. 494 p.

Whitlow, T. H. and R. W. Harris. 1979. Flood tolerance in plants: A state-of-the-art review. Technical Report E-79-2. United States Army Engineer Waterways Experiment Station, Vicksburg, MS.

Wilkin, D. C. and S. J. Hebel. 1982. Erosion, redeposition, and delivery of sediment to midwestern streams. *Wat. Resour. Res.* 18:1278–1282.

Worsley, T. R. and J. M. Dennison. 1973. Sedimentology of White Creek Delta in Watts Bar Lake, Tennessee. Pages 360–375 in W. C. Ackermann, G. F. White, and E. B. Worthington, eds. Man-made lakes: Their problems and environmental effects. Am. Geophys. Union Monograph.

CHAPTER 4

Dissolved Oxygen Dynamics

THOMAS M. COLE AND HERBERT H. HANNAN

The dynamics of dissolved oxygen (DO) in all types of impoundments are based on the same limnological principles and mechanisms described in limnology texts and other literature. Although the generalized profiles of DO for lakes are also found in reservoirs, the overall development and interpretation of these patterns for reservoirs can be different.

Reservoirs have been classified as mainstream (run-of-the-river), transitional, and deep-storage. Although each of these reservoir types is generally considered limnologically different, the limnological characteristics of each type are evident along the main axis of a long deep-storage reservoir. From the shallow riverine reach to the deepest area near the dam, the deep-storage reservoir displays a continuum of all reservoir types.

This chapter first presents the factors affecting DO in reservoirs and then examines the DO dynamics of a generalized, long deep-storage reservoir. Variations in this generalized longitudinal pattern are elucidated. Emphasis is placed on DO dynamics following vernal thermal stratification.

FACTORS AFFECTING THE DISTRIBUTION OF DISSOLVED OXYGEN IN RESERVOIRS

Temperature

Because of the increasing solubility of oxygen with decreasing temperature, colder, more dense water in the deeper layers of a reservoir during summer stratification would have a higher concentration of DO than the warmer, upper layers if DO concentrations were due simply to physical means (Ruttner 1963). This is demonstrated in the well-known orthograde oxygen curve common to oligotrophic lakes. The actual hypolimnetic temperature and DO concentration of a reservoir varies annually from summer to summer and depends largely on the temperature of the water at the onset of vernal stratification. These variations in temperature and DO are more pronounced in reservoirs than in natural lakes due to the effects of inflow and outflow on hypolimnetic temperatures.

The effect of temperature on oxygen concentrations has been demonstrated in many field studies. The apparent eutrophication of Lake Erie as evidenced by an increase in hypolimnetic anoxia over a 30-year period was due in part to an increase in temperature in the hypolimnion during the same period (Charlton 1980). The initial development of the anoxic zone in the upreservoir end of a deep-storage reservoir was attributed in part to the higher temperatures of the hypolimnion when compared to hypolimnetic temperatures downreservoir (Haberle 1981). This relationship was also observed by Weibe (1939a) in his classic paper on Norris Reservoir. Temperature was also the controlling factor in sediment oxygen uptake in five Swedish lakes (Graneli 1978).

Reservoirs that receive heated discharges from steam-generated electrical plants have considerably higher temperatures, which reduce the oxygen saturation values, increase the metabolic activity of organisms, and increase the biochemical oxygen demand (Krenkel et al. 1968). These reservoirs often have longitudinal gradients of decreasing temperature and increasing DO from the discharge outlet to areas within the reservoir.

The temperature pattern of a reservoir is influenced by flow that alters the hypolimnetic temperature during summer stratification. During periods of high outflow in a bottom-release reservoir, the coldest hypolimnetic waters the length of the reservoir are often discharged and replaced by the warmer layers from above. Thus, hypolimnetic temperatures will be greater during years of high outflow than during years of low outflow. In some instances the temperature differences between different years may be as great as 10°C, which would theoretically double the rate of respiration (Haberle 1981). This results in a greater oxygen uptake in the hypolimnion during a year of high flow. Large floods may weaken thermal stratification the length of a reservoir (Wiedenfeld 1980) and have been shown to initiate fall overturn (Ebel and Koski 1968, Raheja 1973).

Flow

The riverine zone of a reservoir receives incoming water from the parent river, which affects the distribution and concentration of DO within all zones of a reservoir. The inflow often forms density currents (see Chapter 2) that can greatly alter the existing DO regimen depending upon their direction of flow and the level at which they move through the reservoir.

Density currents influence the concentration of DO in the epi-, meta-, and hypolimnion. Interflows of water low in DO have been shown to be the cause of metalimnetic oxygen minima in several reservoirs (Weibe 1939a, 1939b, 1940, Dendy 1945). Interflows of oxygenated water often become deoxygenated (Soltero et al. 1974b, Wunderlich 1971), and the resulting metalimnetic oxygen minimum may extend the length of the reservoir. This frequently occurs during flood events (Wiedenfeld 1980). Interflows have also been shown to increase the DO concentrations within a reservoir (Lyman 1944, Hrbacek et al. 1966, Rettig 1980). If the inflowing water has sufficient density it will enter the reservoir as an underflow and may remain on the bottom the length of the reservoir. These underflows can transport water with a higher oxygen content into the hypolimnion (Wiebe 1938, Hrbacek et al. 1966, Eley 1967). This often occurs when reservoirs are located in a series in which the cold hypolimnetic releases from an upstream reservoir enter a downstream reservoir (Dendy and Stroud 1949, Larson 1980). Overflows have very little direct observable effect on DO concentrations in a reservoir because the epilimnion is usually saturated with oxygen from the atmosphere and photosynthesis.

The volume of inflow is often important in determining the DO regimen of a reservoir. Hypolimnetic anoxia that forms in the riverine and transition zones of deep-storage reservoirs following vernal stratification may be dissipated by large inflows (Latif 1973, Leentvaar 1973, Wiedenfeld 1980, Johnson and Page 1980) and reformed following periods of low flow during summer stratification (Haberle 1981). Periods of low flow enable the anoxic zone to develop farther upstream than during normal flow (Haberle 1981). During a period of extremely low flow, the anoxic zone may extend up the riverine reach to the parent river (Hall, Bynum, and Caldwell, unpublished data).

Outlet location is also important in determining the DO distribution within a reservoir. Bottom-release reservoirs usually have a substantially greater portion of hypolimnetic water with DO values above four mg l^{-1}

than reservoirs with surface-level outlets (Tenant et al. 1967, Stroud and Martin 1973). Reservoirs with surface withdrawal increase the residence time of the hypolimnetic waters, which allows oxidative processes to cause greater deoxygenation in the hypolimnion (Stroud and Martin 1973). Dissolved nutrients are also released in the outflow from bottom-release reservoirs, which can result in a decrease in nutrients and a subsequent decrease in primary production, thereby reducing the oxygen demand in the lacustrine zone.

Midlevel withdrawal has essentially the same effect on the overall DO content of the hypolimnion as bottom withdrawal (Stroud and Martin 1973). Midlevel withdrawal currents, however, can cause the anoxic zone to move away from the reservoir bottom to form an oxygen minimum near the outlet with an increase in DO below the withdrawal zone (Ebel and Koski 1968). Midlevel withdrawal currents may have sufficient strength to prevent conventive mixing during the fall and winter (Johnson and Page 1980). As a consequence, the hypolimnion can become low in DO during the winter months.

Morphology

Two lakes with the same trophogenic zones and primary production rates but with different hypolimnetic volumes have different DO concentrations in the hypolimnion (Wetzel 1975). This relationship of hypolimnetic DO and hypolimnetic volume is applicable to reservoirs. The volume of the hypolimnion and the total amount of hypolimnetic DO in a typical reservoir increase from the riverine reach downreservoir to the dam. A decrease in hypolimnetic DO concentrations would, therefore, first occur upreservoir, where the hypolimnion has a relatively smaller amount of DO, and then develop downreservoir with time.

Weibe (1941) noted the importance of morphology on the distribution of DO in his early work with density currents in Norris Reservoir. He predicted that reservoirs that are relatively deep, long, and narrow favor the development of metalimnetic oxygen minima and subsequently observed them in Hiwassee Reservoir and Lake Herrington, which had morphologies similar to Norris Reservoir. This interaction of morphology and flow dynamics, along with temperature, results in several mechanisms that enhance the development of metalimnetic oxygen minima in reservoirs. These mechanisms will be discussed later in the chapter.

Allochthonous Inputs

The parent river supplying a reservoir originates from surface runoff, subsurface seepage, and point sources within the drainage basin. The river water from a particular region within the drainage basin often has its own chemical identity that it either retains or loses as it enters and proceeds through the reservoir. Often it retains its identity and forms cells or layers within the reservoir that are frequently laden with organic matter and nutrients. The organic matter usually results in an oxygen deficiency within the cell or layer, whereas the increase in nutrients often results in a phytoplankton bloom, causing oxygen saturation. These effects are most obvious following large spate events. Water from a 100-year flood entered and moved the length of a deep-storage reservoir as an interflow during the period of thermal stratification (Wiedenfeld 1980). The interflow retained its identity, became anoxic, forming a metalimnetic oxygen minimum in the lacustrine reach, and lowered as water was released from the hypolimnion.

The parent river of a reservoir contains a stream load that is dropped, forming a sedimentation zone, usually within the transition zone. Organic matter coming into a reservoir with inflow does not immediately settle out in the sedimentation zone, but due to its small size and weight, will travel to the downreservoir end of the main sedimentation zone (Frink 1969, Hyne 1978). This organic matter has considerable DO demand, and where it is deposited is usually the area where the anoxic zone first develops in a reservoir.

Photosynthesis and Respiration

Phytoplankton has been shown to be a major contributor of DO to reservoirs. Although important, the contribution of aquatic macrophytes and periphyton to the DO budget in reservoirs has yet to be determined.

Photosynthesis is generally responsible for the commonly observed oxygen pulse in the epilimnion of reservoirs. A diel pulse generally increases from a dawn or postdawn low to a high in the late afternoon and then steadily decreases throughout the night due to continuing demands of community respiration. A diel DO pulse is more common in the riverine reach than in the lacustrine reach of a reservoir, except in large coves, where littoral zone plants and phytoplankton blooms are often abundant.

Metalimnetic oxygen maxima and minima occur in the lacustrine zone of reservoirs in response to DO production and consumption, respectively, by phytoplankton. In addition, metalimnetic oxygen minima have been attributed to bacterial (Kusnetzov and Karsinken 1931, Drury and Gearheart 1975) and zooplankton (Shapiro 1960, Baker et al. 1977) respiration while metalimnetic oxygen maxima have also been attributed to littoral zone macrophytes (Wetzel 1975).

Hypolimnia deficient in DO are common to reservoirs. The importance of bacterial respiration and its effects on DO concentrations at the sediment-water interface is well known, but recent studies have shown the importance of bacterial respiration in the DO dynamics of the upper layers of the hypolimnion (Lund et al. 1963, Lasenby 1975, Cangialosi 1976, Lepak 1976).

Wind

The primary effect of wind on the DO distribution in reservoirs is through wind-induced mixing, which moves water in the lower layers to the surface. This causes a gain in oxygen by helping to maintain the partial pressure differential necessary to sustain oxgyen diffusion into the water, or a loss of oxygen by bringing supersaturated waters to the surface, where oxygen is lost to the atmosphere. In either case wind mixing aids in maintaining relatively uniform oxygen concentrations in the epilimnion during stratification and throughout the entire water column during and after overturn. During a period of a few days with little or no wind it is common to find an increase in oxygen due to photosynthesis associated with phytoplankton blooms. Subsequent winds will spread the plankton along the fetch and cause cells of higher DO (Thornton, personal communication). The die-off and decomposition of these blooms often create a DO deficiency far removed from the initial bloom (Wegner, personal communication).

The most striking effect of wind on DO distribution can occur during early fall when high winds associated with cold fronts can bring about complete overturn in one day. Beadle (1974) presents evidence that wind is the primary factor in determining the permanency of stratification and thus the seasonal distribution of DO in subtropical lakes. He also points out that because of the slight temperature differences that often occur between surface and bottom waters, wind-induced evaporative cooling at the surface may be sufficient to reduce the thermal stability of the water and favor deep mixing by winds.

The effectiveness of wind in promoting mixing is influenced by the strength of the wind, its direction in relation to the orientation of the reservoir, the surrounding topography that determines the exposure of the reservoir to the wind, the fetch, the depth of the water, and the thermal stability of the reservoir. For reservoirs, most of these factors change from the riverine zone to the lacustrine zone, so that the effects of wind on DO concentrations change longitudinally. Wind mixing is generally greatest in the lacustrine zone where the fetch and exposure to wind are the greatest. Oxygen concentrations change less over a greater depth in the epilimnion of the lacustrine zone than in the riverine zone.

A GENERAL PATTERN FOR HYPOLIMNETIC OXYGEN DEPLETION IN RESERVOIRS

The pattern of hypolimnetic DO depletion is highly variable both spatially and temporally for any one reservoir and among reservoirs. There is, however, a general pattern of hypolimnetic DO depletion that has been found to be common in both oligotrophic and eutrophic reservoirs (Table 4.1). Following the onset of vernal stratification, the anoxic zone first develops in the thalweg at the downreservoir end of the active sedimentation zone and then progresses both upreservoir and downreservoir. This longitudinal development continues throughout summer stratification until the anoxic zone reaches the parent free-flowing river at one end and the dam at the other end or until fall overturn. Concurrently, the anoxic zone develops upward and laterally out of the thalweg until it either reaches a relatively stationary metalimnion, intergrades with a deepening metalimnion, which is often low in DO, or is dissipated by fall overturn. The hypolimnetic anoxic zone often develops in the thalweg the length of the reservoir before it develops up to the metalimnion (Haberle 1981). The time of development ranges from a few days for shallow eutrophic reservoirs to months for long reservoirs with an oligotrophic lacustrine zone (Hannan 1979).

This generalized pattern of hypolimnetic DO depletion results from an interaction of flow and morphology, which in turn affects sedimentation, primary production, DO available to meet hypolimnetic oxygen demands, and the hypolimnetic temperature regime (Figure 4.1). Oxygen depletion first occurs in the transition zone at the downreservoir end of the sedimentation zone where there is a large input of small-sized allochthonous organic matter and a high rate of oxygen uptake. A decrease in velocity

Table 4.1 Reservoirs Exhibiting a Downreservoir Development of a Hypolimnetic anoxic zone

Reservoir	Study
Norris Reservoir, Tennessee	Weibe (1938); Dendy (1945)
Elephant Butte Reservoir, New Mexico	Ellis (1940)
Cherokee Reservoir, Tennessee	Lyman (1944); Iwanski et al. (1979); Gordan and Nichols (1977)
Williamette River, Oregon	Fish and Wagner (1950)
Lake Lanier, Washington	Vanderhoof (1965)
Slapy Reservoir, Czechoslovokia	Hrbacek and Straskraba (1966)
Fontana Reserovir, North Carolina	Louder and Baker (1966)
Klicava Reservoir, Czechoslovokia	Fiala (1966)
Brownlee Reservoir, Oregon-Idaho	Ebel and Koski (1968)
Bighorn Lake, Montana–Wyoming	Soltero et al. (1974a)
Long Lake, Washington	Soltero et al. (1974a, 1975)
DeGray Reservoir, Arkansas	Nix (1974)
Center Hill Reservoir, Tennessee	Gnilka (1975)
Lake Kariba, Zambia–Rhodesia	Bowmaker (1976)
Canyon Reservoir, Texas	Hannan et al. (1979); Wiedenfeld (1980); Haberle (1981); Hall, Caldwell, and Bynum (unpublished data, 1981); Bira (unpublished, 1982)
Lake Livingston, Texas	Rawson (1979)
Shasta Lake, California	Rettig (1980)
Lake Norman, South Carolina	Foris (pers. comm., 1981)
Beaver Lake, Arkansas	Brown (pers. comm., 1981)

in this zone along with an increase in light penetration due to sedimention allows the buildup of phytoplankton populations which also contribute organic matter to the hypolimnion. Primary production in the transition zone is maintained at a high level since the phytoplankton community receives a continuous supply of nutrients from the inflowing water and the sediment.

How rapidly hypolimnetic deoxygenation occurs depends on the total amount of DO available in the hypolimnion and the temperature at which oxidation of organic matter occurs. The hypolimnion of reservoirs has a smaller volume in the transition zone as compared to the lacustrine zone.

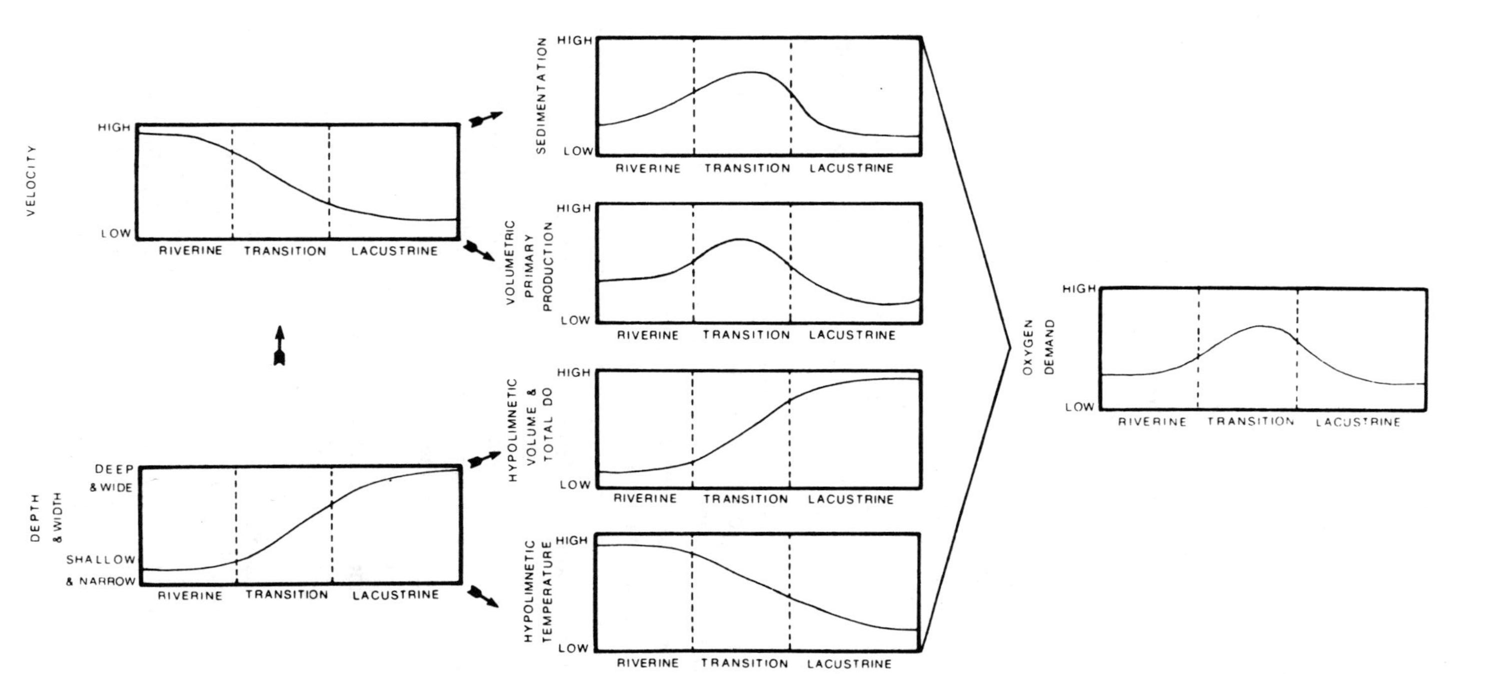

Figure 4.1 Relationships of velocity and depth to sedimentation, areal primary production, hypolimnetic volume and total dissolved oxygen, and hypolimnetic temperature in determining oxygen demand the length of a reservoir.

This results in a lower total volume of DO to meet hypolimnetic oxygen demands. The morphology of reservoirs also results in relatively higher overall hypolimnetic temperatures in the transition zone than in the lacustrine zone, which in turn cause higher respiration rates. The higher temperatures also make less DO available in the hypolimnion of the transition zone due to the relationship between oxygen solubility and temperature. The result of these factors is that the highest biochemical oxygen demand occurs in the hypolimnion of the transition zone, causing the anoxic zone to first form in this region.

The time of onset and duration of hypolimnetic anoxic conditions during summer stratification varies by as much as two months and is the result of annual differences in temperature, flow, and wind action (Wegner, personal communication). Warm, calm, low inflow years allow the reservoir to stratify early during the spring, which allows anoxic conditions to develop early in the summer. Conversely, cold, windy, high inflow years may delay thermal stratification, which delays hypolimnetic anoxic conditions until late in the summer. Obviously, the onset and duration of hypolimnetic anoxia depend on meteorological factors controlling the onset of vernal stratification as well as fall overturn. The annual differences in the onset and duration of hypolimnetic anoxic conditions are more pronounced in regions where there are large annual differences in meteorological conditions.

The site at which the anoxic zone initially forms and its upstream and downstream boundaries are determined primarily by flow (Figure 4.2). In order for anoxia to develop, the bottom layers of water must be isolated from mechanisms of reaeration. A period of high inflow extends reaeration farther downreservoir and also shifts the sedimentation zone downreservoir. This results in the initial site of the anoxic zone shifting downreservoir. A year of lower inflow results in the sedimentation zone forming farther upreservoir and allows stratification to extend farther upreservoir, thereby shifting the initial formation of the anoxic zone in that direction.

Once formed, the anoxic zone develops longitudinally both upreservoir and downreservoir. The upreservoir development is influenced primarily by inflow from the parent river. As spring inflows subside, thermal stratification extends farther up the riverine reach, which isolates the bottom waters from reaeration and allows anoxic conditions to develop. Any flood event that breaks down thermal stratification and allows mixing

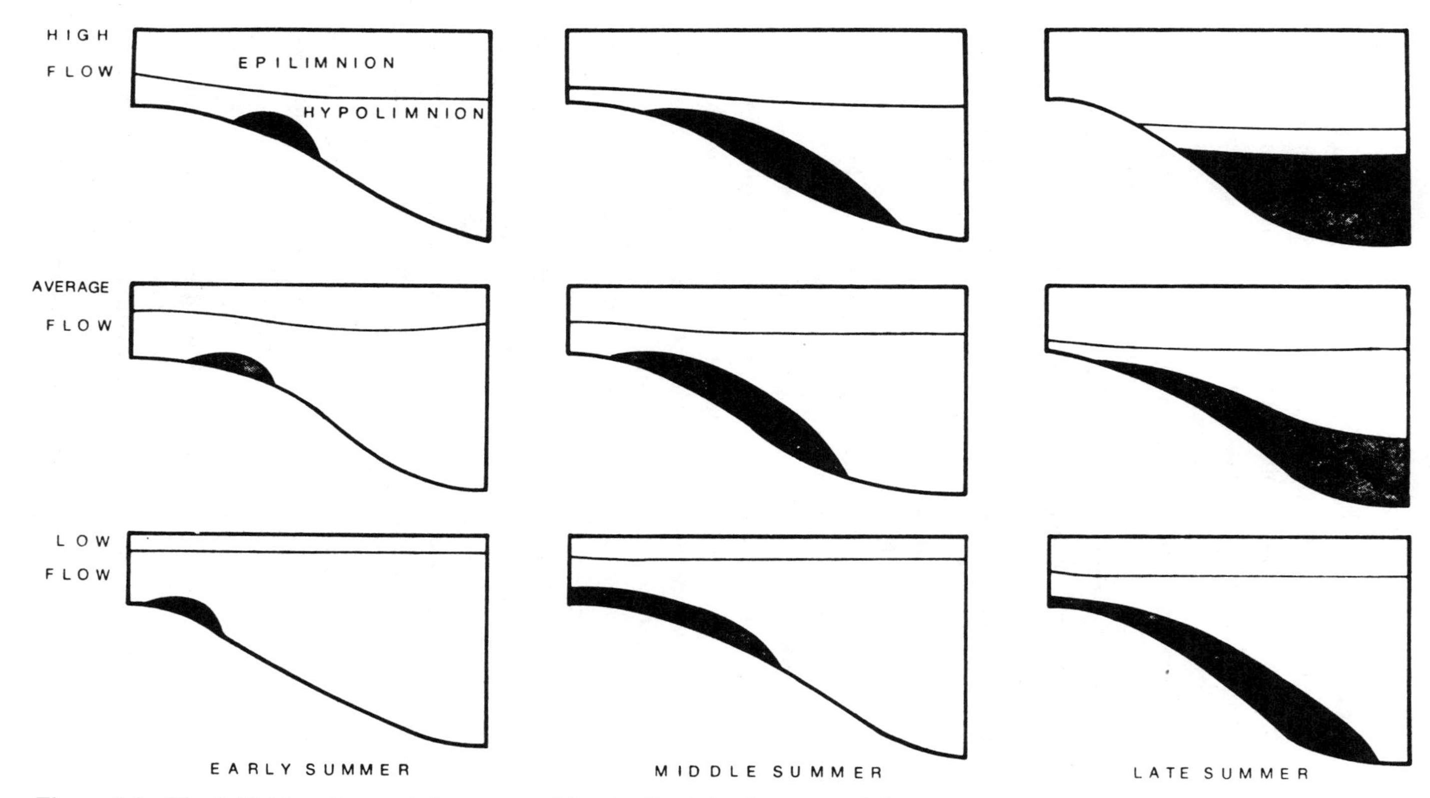

Figure 4.2 The initial location and the extent of longitudinal development of the anoxic zone during years of high, average, and low flow in a deep-storage reservoir.

to take place will dissipate the anoxic zone, the extent of which depends on the volume of inflow. As a flood subsides, anoxic conditions often rapidly redevelop. The upreservoir development of the anoxic zone can also be influenced by the presence of reverse density currents in the hypolimnion moving anoxic waters upstream (Lawrence 1967).

The anoxic zone develops downreservoir slower and with less fluctuation than upreservoir. The anoxic zone develops downreservoir faster in a reservoir with a hypolimnetic outlet than in one with an epilimnetic outlet. Moreover, the anoxic zone develops downreservoir faster in a reservoir with a hypolimnetic outlet during a year of high spring and summer rains than during either a normal or drought year due to an increase in withdrawal currents. This is especially true in reservoirs that are maintained at a constant water level.

Withdrawal also accelerates the downreservoir movement of the anoxic zone by influencing the temperature and volume of the hypolimnion. Hypolimnetic temperatures are higher during years of high flow since more of the hypolimnetic waters are withdrawn and replaced by the warmer waters from the layers above. The higher temperatures cause an increase in the rate of oxygen uptake and a decrease in oxygen solubility, accelerating the downreservoir development of the anoxic zone. Higher withdrawal rates also decrease the volume of the hypolimnion. This decreases the total amount of DO available to meet hypolimnetic oxygen demands, resulting in more rapid downreservoir development of the anoxic zone. A year of high flow also decreases the density and viscosity differences in the metalimnion and allows more allochthonous organic matter to reach the hypolimnion, thereby accelerating deoxygenation. A year of low flow has an opposite effect that results in a smaller volume of anoxic water in the hypolimnion and a potential later arrival of the anoxic zone at the dam.

Hypolimnetic anoxic development in the lacustrine zone is the result of deoxygenation occurring from as many as three regions. Hypolimnetic anoxia develops from the thalweg up to the metalimnion or as a result of the anoxic water moving downreservoir (Figure 4.2) intergrading with oxygen deficient water from both the bottom and the upper hypolimnetic-metalimnetic region (Figure 4.3). This latter pattern of hypolimnetic anoxic development is contingent upon the formation of metalimnetic oxygen minima, which are common in reservoirs. Hypolimnetic anoxia can develop in a cove with a large epilimnetic-hypolimnetic ratio before it develops in the thalweg. In this case the anoxic water from the cove may develop to the thalweg of the reservoir prior to anoxia forming in the thalweg.

Although this general pattern of DO development was formulated primarily from studies on deep-storage reservoirs, it has the characteristics of a continum of reservoir types from shallow eutrophic reservoirs with a large allochthonous input and small downreservoir hypolimnetic volume to long reservoirs with an oligotrophic lacustrine reach with a small to large allochthonous input and a large downreservoir hypolimnetic volume (Hannan 1979). For example, the hypolimnetic DO dynamics of a shallow eutrophic reservoir (Young et al. 1972) are similar to those of the riverine reach of a deep-storage reservoir (Haberle 1981). Both systems are primarily controlled by inflow, which causes variable DO conditions. If an anoxic zone does develop it should first form rapidly near the dam or in conjunction with the sedimentation zone and develop upreservoir during periods of low flow and be dissipated during periods of high flow. It is common for an anoxic zone formed in both of these systems to be dissipated and reformed within a few days.

This generalized pattern of hypolimnetic DO development shows that the classic clinograde and orthograde oxygen curves frequently used to indicate trophic conditions in lakes are not applicable to reservoirs. This is especially true for deep-storage reservoirs. As previously stated, the hypolimnion first becomes anoxic in the transition zone. This results in a continuuum of DO profiles along the length of a reservoir from shallow clinograde curves in the transition zone to deep orthograde curves at the dam. As hypolimnetic anoxic conditions develop downreservoir as summer progresses, there is a concomitant clinograde curve formation. Therefore, a lacustrine area that is often characteristically oligotrophic has a DO profile that is characteristically eutrophic by the middle or at the end of summer stratification.

This hypolimnetic oxygen depletion pattern has been synthesized from field studies on numerous reservoirs and from individual reservoirs studied over a period of years. Variations in this pattern should be expected. Often one or more of the principal factors affecting DO distribution may dominate the development of this pattern during any one year. It is hoped that this general pattern permits one to understand these special cases. The purpose of the next section is to identify the more common variations and their causal factors.

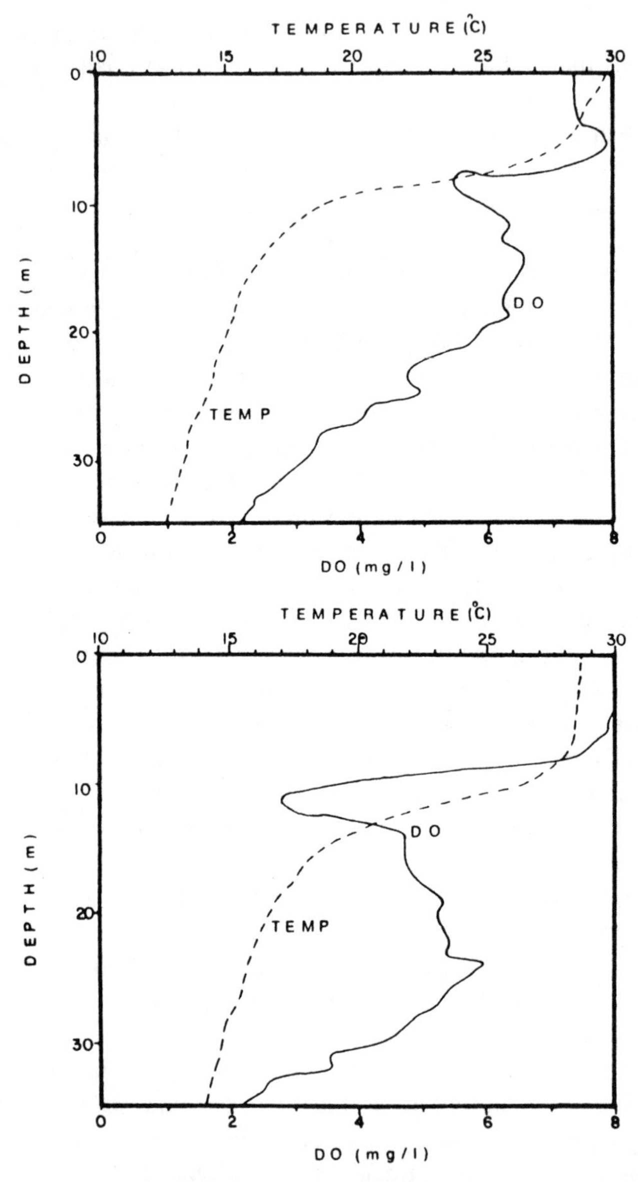

Figure 4.3 The intergradation of oxygen-deficient water from the metalimnion and bottom of the lacustrine zone during the formation of an anoxic hypolimnion in Canyon Reservoir (Hall, Bynum, and Caldwell unpublished data).

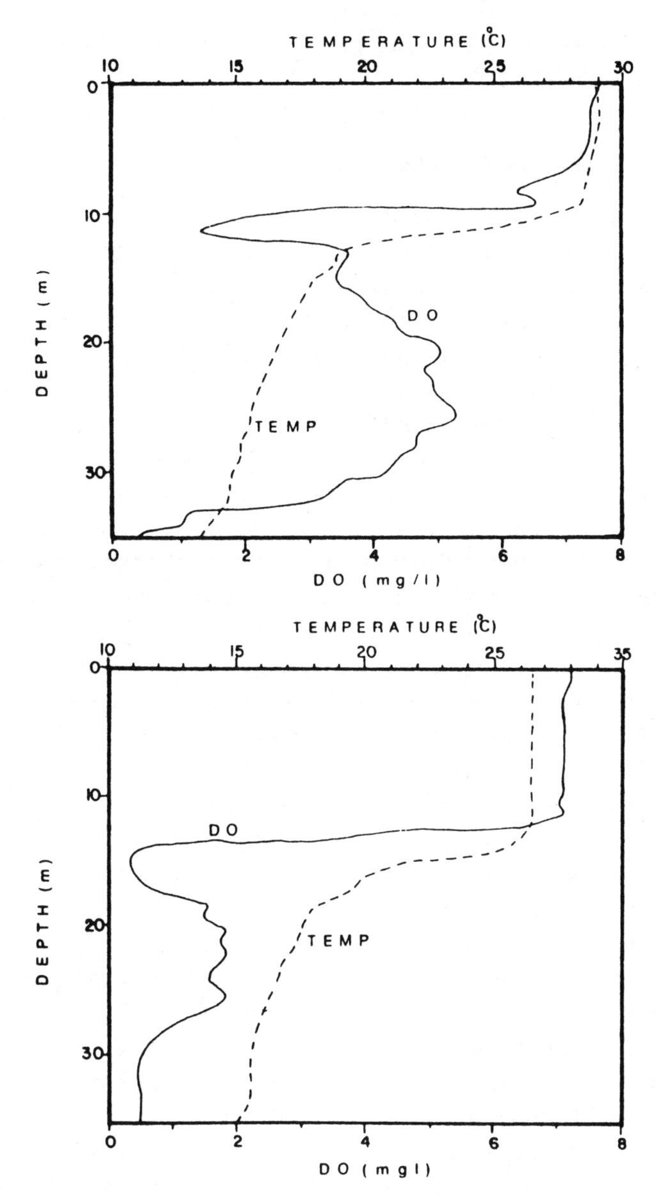

Figure 4.3 *continued*

VARIATIONS IN THE GENERAL LONGITUDINAL PATTERN OF OXYGEN DEPLETION

Metalimnetic Oxygen Minima

Metalimnetic oxygen minima are a common occurrence in reservoirs (Table 4.2). Several processes have been proposed to explain the cause of metalimnetic oxygen minima in lakes and reservoirs (Table 4.3). The decomposition of seston trapped in the metalimnion due to the sharp increases in density and viscosity associated with this layer has been proposed to explain the presence of metalimnetic oxygen minima (Birge and Juday 1911). Both bacteria (Drury and Gearheart 1975, Gordon and Skelton 1977) and zooplankton (Shapiro 1960, Baker et al. 1977) may be responsible for the respiration of the trapped seston. The degree of thermal stratification may be an important factor in determining the relative importance of the decomposition of seston in the development of metalimnetic oxygen minima. In Canyon Reservoir minima occur more frequently and are more pronounced during years of low flow than during years of high flow. Apparently, the increased density and viscosity gradients in the metalimnion during low-flow years result in a greater amount of seston being trapped in the metalimnion.

Morphology also contributes to the development of metalimnetic oxygen minima. If the metalimnion coincides with a portion of the bottom that has a minimal slope, then a greater area of metalimnetic water will come into contact with the sediments, resulting in a greater uptake of oxygen in this stratum. This "shelf effect" has been proposed to explain metalimnetic oxygen minima in lakes (Hutchinson 1957). A common mechanism found to be responsible for metalimnetic oxygen minima in reservoirs is the interflow of water with either a low DO concentration or a high biochemical oxygen demand. Interflows, particularly during flood events, may bring about metalimnetic oxygen minima due to the high oxygen demand associated with the highly turbid water (Lyman 1944). Oxygen depletion caused by a 100-year flood was observed in an interflow in Canyon Reservoir (Wiedenfeld 1980). A metalimnetic oxygen minimum was quickly established following the flood and persisted until fall overturn.

These metalimnetic oxygen minima associated with interflows often form or persist because they are usually located below the euphotic zone, where the nutrients available in the interflow cannot be utilized by the

Table 4.2 Reservoirs in which metalimnetic oxygen minima have been observed

Reservoir	Study
Norris Reservoir, Tennessee	Weibe (1938, 1939a, 1940)
Elephant Butte Reservoir, New Mexico	Ellis (1940)
Lake Herrington, Kentucky	Weibe (1941)
Hiwassee Reservoir, North Carolina	Weibe (1941)
Cherokee Reservoir, Tennessee	Lyman (1944)
Calderwood Reservoir, Tennessee	Dendy and Stroud (1949)
Fontana Reservoir, North Carolina	Dendy and Stroud (1949); Wunderlich (1971)
Slapy Reservoir, Czechoslovakia	Hrbacek and Straskraba (1966), Fiala (1966)
Boone Reservoir, Tennessee	Churchill and Nicholas (1967)
Bull Shoals Reservoir	Mullan et al. (1970)
Lake Mead, Arizona–Nevada	Hoffman and Jones (1973), Baker et al. (1977)
Bighorn Lake, Montana	Soltero et al. (1974)
Center Hill Reservoir, Tennessee	Gnilka (1975), Gordon and Skelton (1975), Gordon and Morris (1979)
Beaver Lake, Arkansas	Drury and Gearheart (1975)
Dworshak Reservoir, Idaho	Falter (1976)
South Holston Reservoir, Tennessee	Gordon and Skelton (1977)
Tims Ford Reservoir, Tennessee	Gordon and Skelton (1977)
Canyon Reservoir, Texas	Wiedenfeld (1978), Hannan (1979), Hall (1980), Bira (1982)
Flaming Gorge Reservoir, Wyoming	Bolke (1979)
Lake Powell, Utah–Arizona	Johnson and Page (1980)
Shasta Lake, California	Rettig (1980)
Deer Creek Reservoir, Utah	Wegner (pers. comm., 1982)
DeGray Reservoir, Arkansas	Thornton (pers. comm., 1982)
Lost Creek Reservoir, Utah	Wegner (pers. comm., 1982)
Scofield Reservoir, Utah	Wegner (pers. comm., 1982)

phytoplankton. They occur most frequently in the lacustrine zone of bottom-draining deep-storage reservoirs in middle or later summer after the metalimnion has been lowered out of the euphotic zone due to hypolimnetic withdrawal. The magnitude of these minima caused by

Table 4.3 Possible Mechanisms Responsible for metalimnetic oxygen minima in lakes and reservoirs

	Observed in	
Mechanism	Lake	Reservoir
Decomposition of seston	Shapiro (1960)	Drury and Gearheart (1975)
Shelf effect	Hutchinson (1957), Shapiro (1960)	
Interflow		Lyman (1944), Wiedenfeld (1980)
Withdrawl currents		Ebel and Koski (1968)
Cove effect		Johnson and Page (1980)
Temperature effect		
Circulation pattern		Wunderlich (1971)

interflows has been attributed to the sequence of elevated spring flows and the onset of thermal stratification (Nix, personal communication). If thermal stratification precedes elevated spring flows, organic loading may enter the metalimnion, causing greater oxygen demand. If high spring flows precede the onset of thermal stratification, much of the allochthonous organic matter is lost to the hypolimnion.

Interflows also bring about the appearance of a metalimnetic oxygen minimum, which is actually a metalimnetic oxygen maximum. If water with a high DO concentration enters a reservoir exhibiting a clinograde DO stratification near the bottom of the metalimnion or slightly below it, then the resulting DO distribution will have the appearance of a negative heterograde curve (Figure 4.4). A metalimnetic oxygen minimum may also result from withdrawal currents caused by a midlevel outlet (Ebel and Koski 1968). These currents pull anoxic water off the bottom from upstream toward the outlet, forming a metalimnetic oxygen minimum, while the more oxygenated water beneath the outlet remains stationary (Figure 4.5). The development of a metalimnetic oxygen minimum can also be influenced by coves or bays (Johnson and Page 1980). Low DO concentrations that develop in the hypolimnion at the head of a bay and extend outward into the metalimnion can contribute to a metalimnetic oxygen minimum in the open waters of a reservoir (Figure 4.6). This

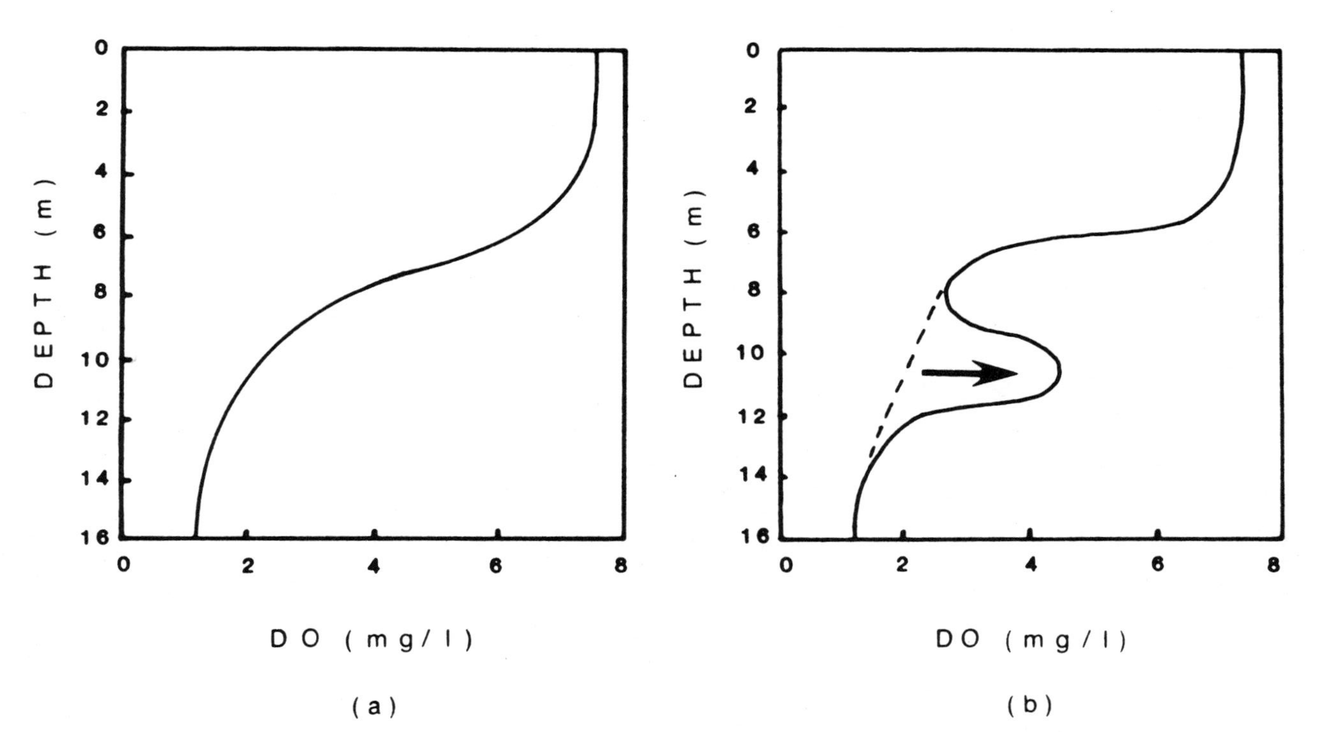

Figure 4.4 The formation of a metalimnetic oxygen minimum caused by an interflow with a high dissolved oxygen content. (*a*) Clinograde oxygen curve. (*b*) Clinograde oxygen curve modified by an interflow of highly oxygenated water.

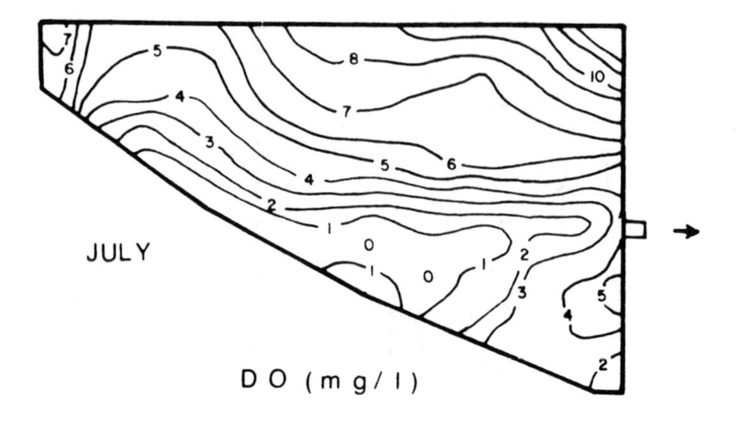

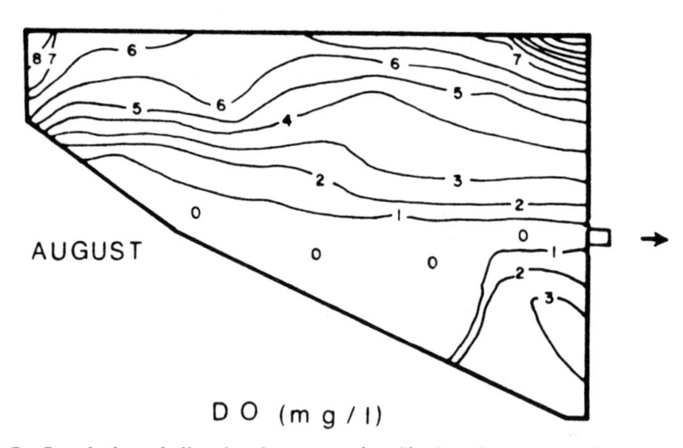

Figure 4.5 Isopleths of dissolved oxygen (mg/l) showing a metalimnetic oxygen minimum caused by currents formed by a midlevel outlet (modified from Ebel and Koski 1968).

pattern has recurred in several different years in numerous bays of Lake Powell.

The relationship between phytosynthetic production of oxygen and the effect of temperature on respiration rates and oxygen solubility often contributes to the formation of metalimnetic oxygen minima. In the upper layers of the epilimnion, phytosynthetic production of oxygen masks the effects of community respiration. This masking effect decreases with

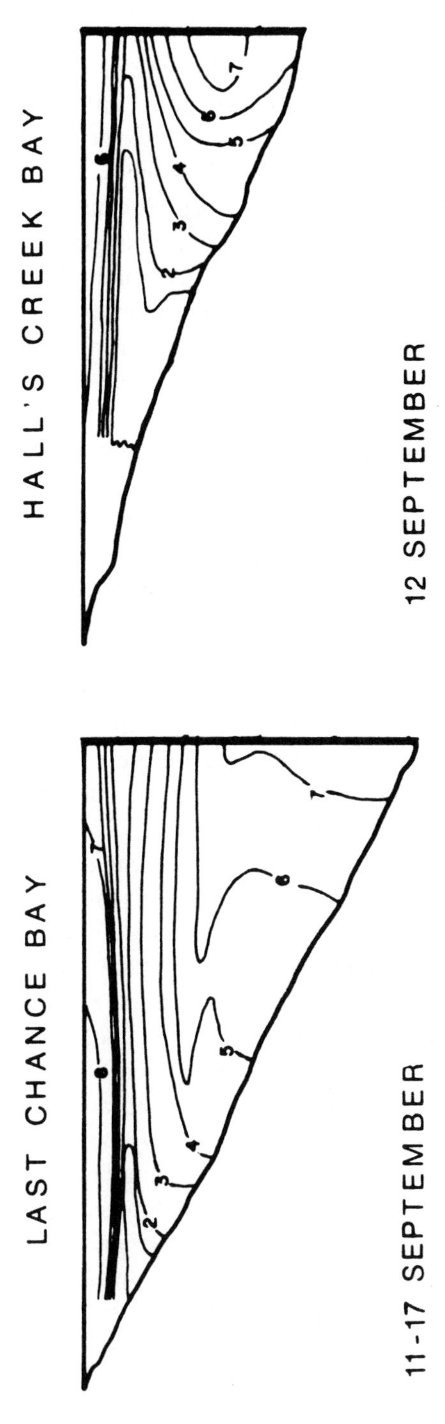

Figure 4.6 Isopleths of dissolved oxygen (mg/l) showing metalimnetic oxygen minima formed at the mouths of coves located along the main axis of Lake Powell (modified from Johnson and Page, 1980).

depth until at the compensation point there is little or no production but considerable consumption of oxygen. Below the metalimnion, respiration rates decrease due to the lower temperature and DO concentrations increase due to the increased solubility of oxygen in water at lower temperatures. The organic matter, which reaches the hypolimnion, will also be more resistant to oxidation since the more easily oxidizable compounds are oxidized in the epilimnion. The overall result will be a negative heterograde curve with an oxygen minimum occuring in the region of the metalimnion (Ruttner 1963). Shapiro (1960) suggests that this is a common mechanism, and the apparent absence of an oxygen minimum in many bodies of water is due to large sampling depth intervals.

The results of two-dimensional computer models suggest another possible mechanism that may be responsible for the development of metalimnetic oxygen minima in reservoirs. Several studies have shown that an oval circulation pattern is common in the epilimnion of reservoirs during summer stratification (Edinger and Buchak 1977, Gordon 1980) (Figure 4.7*a*). This flow pattern moves oxygenated water from the surface to the lower layers and oxygen-deficient water from the lower layers to the surface, which results in oxygenated water surrounding a cell of stagnant water. A DO minimum could develop in the isolated cell of water over time. Oxygen data from Fontana Reservoir (Wunderlich 1971) suggest that a circulation pattern of this type may be influencing the observed minimum (Figure 4.7*b*)

A metalimnetic oxygen minimum is more common in the lacustrine reach of deep-storage reservoirs than in other reaches or shallower impoundments. The absence of a metalimnetic oxygen minimum in transitional type reservoirs and in the transition zone of deep-storage reservoirs is a function of depth. In the transition zone of a deep-storage reservoir, a clinograde oxygen profile (Figure 4.8*a*) forms as a result of anoxic conditions at a depth similar to that where a negative heterograde curve forms in the lacustrine zone (Figures 4.8*b* and 4.8*c*).

In a discussion of the possible mechanisms responsible for metalimnetic oxygen minima in lakes, Wetzel (1975) points out that in most cases several mechanisms contribute to the development of a metalimnetic oxygen minimum, and to single out only one mechanism to explain a minimum is wrong. This is also true for reservoirs, but when the large number of metalimnetic oxygen minima that have been observed in reservoirs is considered, it appears that interflows entering metalimnia

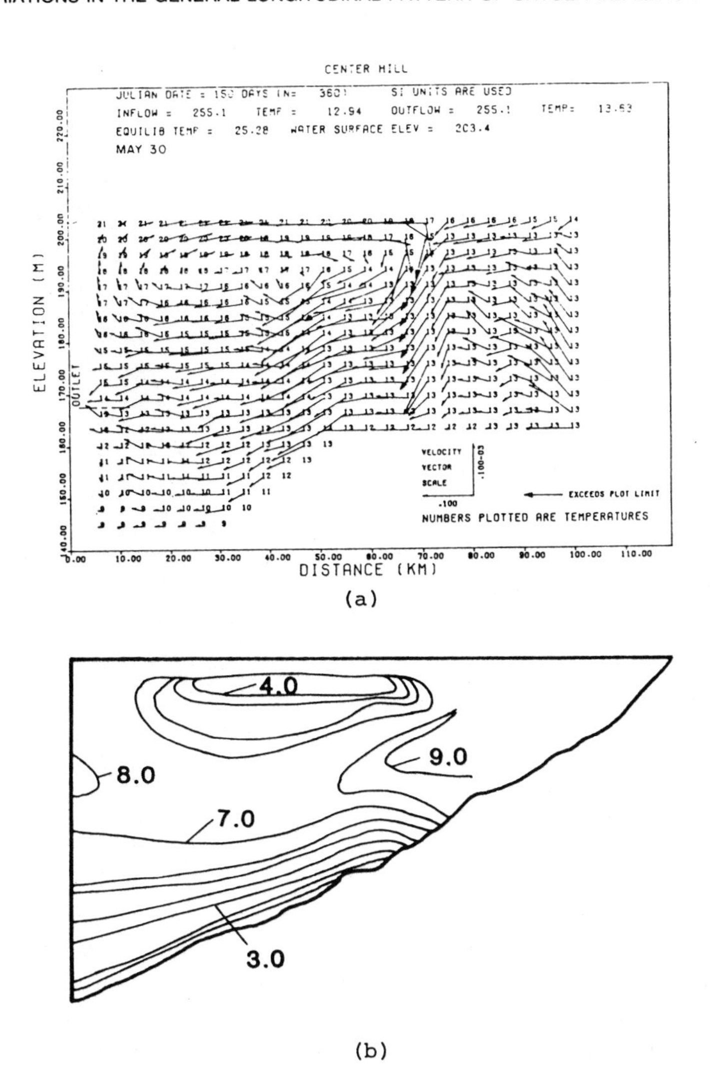

Figure 4.7 (*a*) A computer simulation of the circulation pattern in Center Hill Reservoir resulting from an interflow (from Gordon 1981). (*b*) Isopleths of dissolved oxygen (mg/l) in Fontana Reservoir (modified from Wunderlich 1971).

and metalimnia with decomposing seston being lowered below the euphotic zone are important controlling factors. This could explain why minima are much more common than maxima in reservoirs whereas the opposite is thought to be true in lakes (Wetzel 1975).

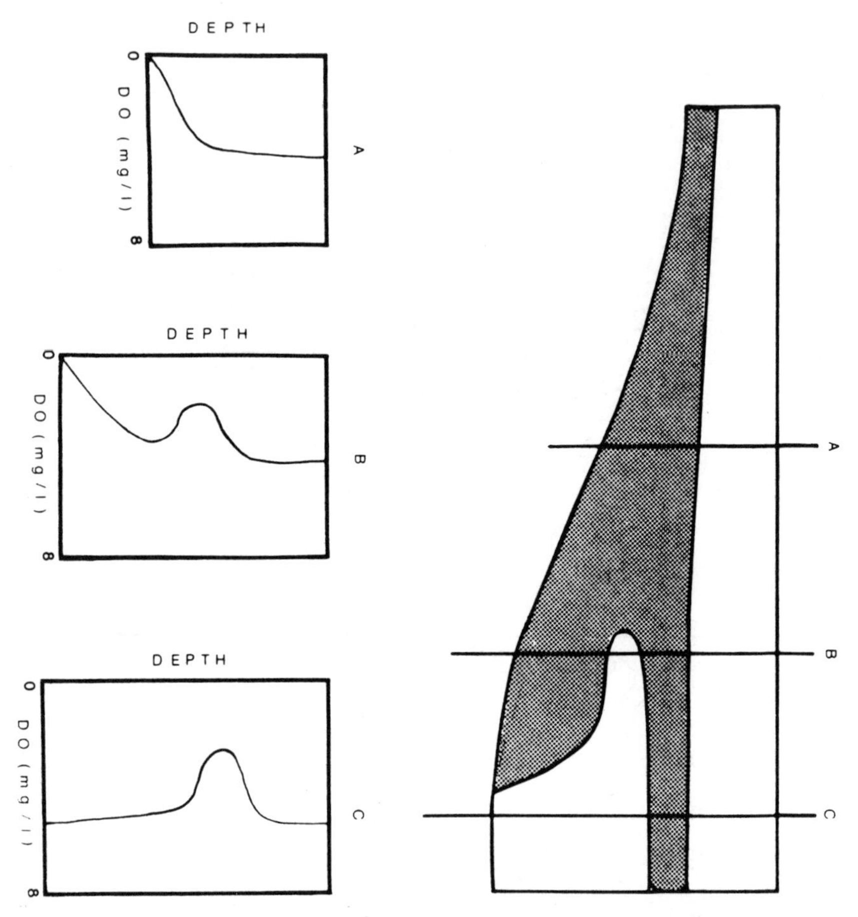

Figure 4.8 A comparison of vertical dissolved oxygen profiles the length of a deep-storage reservoir. Shaded area represents dissolved oxygen less than 4 mg/l. (*a*) Clinograde oxygen curve. (*b*) Metalimnetic oxygen minimum with deoxygenation in the hypolimnion. (*c*) Orthograde oxygen curve with metalimnetic oxygen minimum.

Epilimnetic and Metalimnetic Oxygen Maxima

Epilimnetic oxygen maxima are generally associated with oxygen produced by rooted macrophytes in the littoral zone and phytoplankton blooms in the limnetic zone of a reservoir. Diel flunctuations in DO in both areas often range from zero to over 200% saturation during a calm day. Strong winds often cause a bloom to be dispersed into streaks or bands with a concomitant or subsequent increase in DO. These conditions are prominent in large coves as well as in the open parts of a reservoir.

Metalimnetic oxygen maxima have been less frequently observed in reservoirs than metalimnetic oxygen minima. Presumably the same mechanisms responsible for metalimnetic oxygen maxima in lakes are in effect in reservoirs, the most common type being the increased solubility of oxygen in the colder waters of the metalimnion with a typical clinograde decrease in DO with depth in the hypolimnion. When the euphotic zone extends into the metalimnion, a buildup of a phytoplankton community may result in a metalimnetic oxygen maximum (Hutchinson 1957). A metalimnetic oxygen maximum in reservoirs may also be due to the interflow of more highly oxygenated waters (Hrbacek and Straskraba 1966).

It is possible that a metalimnetic oxygen maximum occurs most frequently during early summer and precedes a metalimnetic oxygen minimum that occurs most frequently later in the summer. A plausible explanation is that a metalimnetic oxygen maximum would be promoted during early summer when the spring interflows are laden with nutrients, thermal stratification has occurred, and the euphotic zone extends into the metalimnion. Subsequent phytoplankton growth in the epilimnion would decrease the depth of the euphotic zone by shading with respiration exceeding photosynthesis and a metalimnetic oxygen minimum forming below the metalimnetic oxygen maximum. If the reservoir is bottom-draining, a metalimnetic oxygen maximum would be lowered out of the euphotic zone as summer progresses, and a metalimnetic oxygen minimum would then develop in middle or late summer. A metalimnetic oxygen maximum would also be less frequent than a metalimnetic oxygen minimum in late summer or early fall as it would be dissipated by the epilimnion deepening with the onset of seasonal cooling.

Oxygen Block

An "oxygen block," a zone of oxygen-deficient water, has been observed in the transition zone of several reservoirs (Ellis 1940, Fish and Wagner 1950, Ebel and Koski 1968). The oxygen block extends from the surface to the bottom and is surrounded on all sides by water containing a higher concentration of DO (Figure 4.9*a*). Dye studies of Eufala Reservoir (Lawrence 1967) suggest that oxygen blocks may be due to a reverse

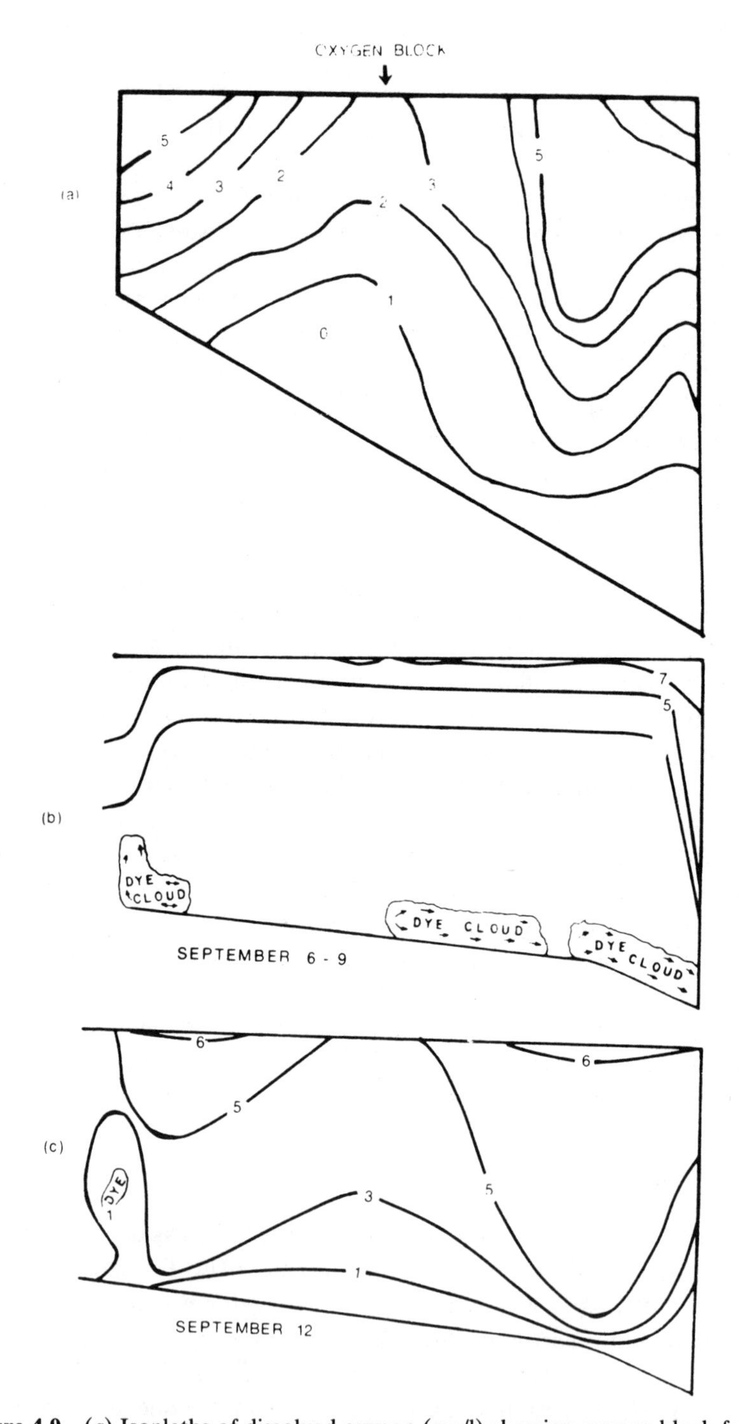

Figure 4.9 (*a*) Isopleths of dissolved oxygen (mg/l) showing oxygen block formation in the transition zone of Brownlee Reservoir (modified from Ebel and Koski 1968). (*b*) Movement of density currents in the transition zone of Eufala Reservoir (modified from Lawrence 1967). (*c*) Isopleths of dissolved oxygen (mg/l) showing oxygen block formation in Eufala Reservoir due to reverse density currents (modified from Lawrence 1967).

density current that encounters the inflowing water and then moves upward, bringing the bottom waters low in DO to the surface (Figures 4.9*b* and 4.9*c*).

These conditions resemble a front that occurs in large lakes and oceans. A front is an aquatic region of high biological activity located between two different aquatic regions. An oxygen block could well be a characteristic of the transition zone that separates the riverine from the lacustrine reach of a reservoir and may easily be overlooked in a study unless comprehensive sampling is undertaken.

Akinetic Spaces

Layers with restricted water motion (akinetic spaces) influence the patterns of DO distribution in reservoirs (Churchill 1958, Goda 1959, Fiala 1966, Ebel and Koski 1968, Haberle 1981). The effects of akinetic spaces on DO concentrations are similar to the effects of stratification in which the hypolimnion becomes isolated from reaeration mechanisms and DO concentrations are lowered. Four possible mechanisms have been observed by which akinetic spaces may form. Midlevel withdrawal can isolate the layers below the outlet to form an akinetic space (Figure 4.10*a*) (Ebel and Koski 1968, Fiala 1966, Johnson and Page 1980). Submerged weirs may also restrict water movement sufficiently to form an akinetic space (Figure 4.10*b*), and in some cases a monimolimnion can form, which is undisturbed even during overturn periods (Fiala 1966). If outlets are located a sufficient distance upstream from the dam, an akinetic space can form between the outlet and the dam (Figure 4.10*c*) (Fiala 1966). Depressions located in the thalweg may contain water that is several degrees colder than the rest of the reservoir, which can result in the formation of an akinetic space (Figure 4.10*d*) (Haberle 1981). Anoxic conditions that develop in these depressions often persist for as much as a month after the rest of the reservoir has overturned.

Meromixis

In his discussion of meromixis in lakes Hutchinson (1957) states that "it is not unlikely that the definition of crenogenic meromixis may have to be widened to include cases where solid salts in the bottom deposits are going into solution or where saline rivers or dense industrial wastes enter the lake at the surface and flow into the monimolimnion as density

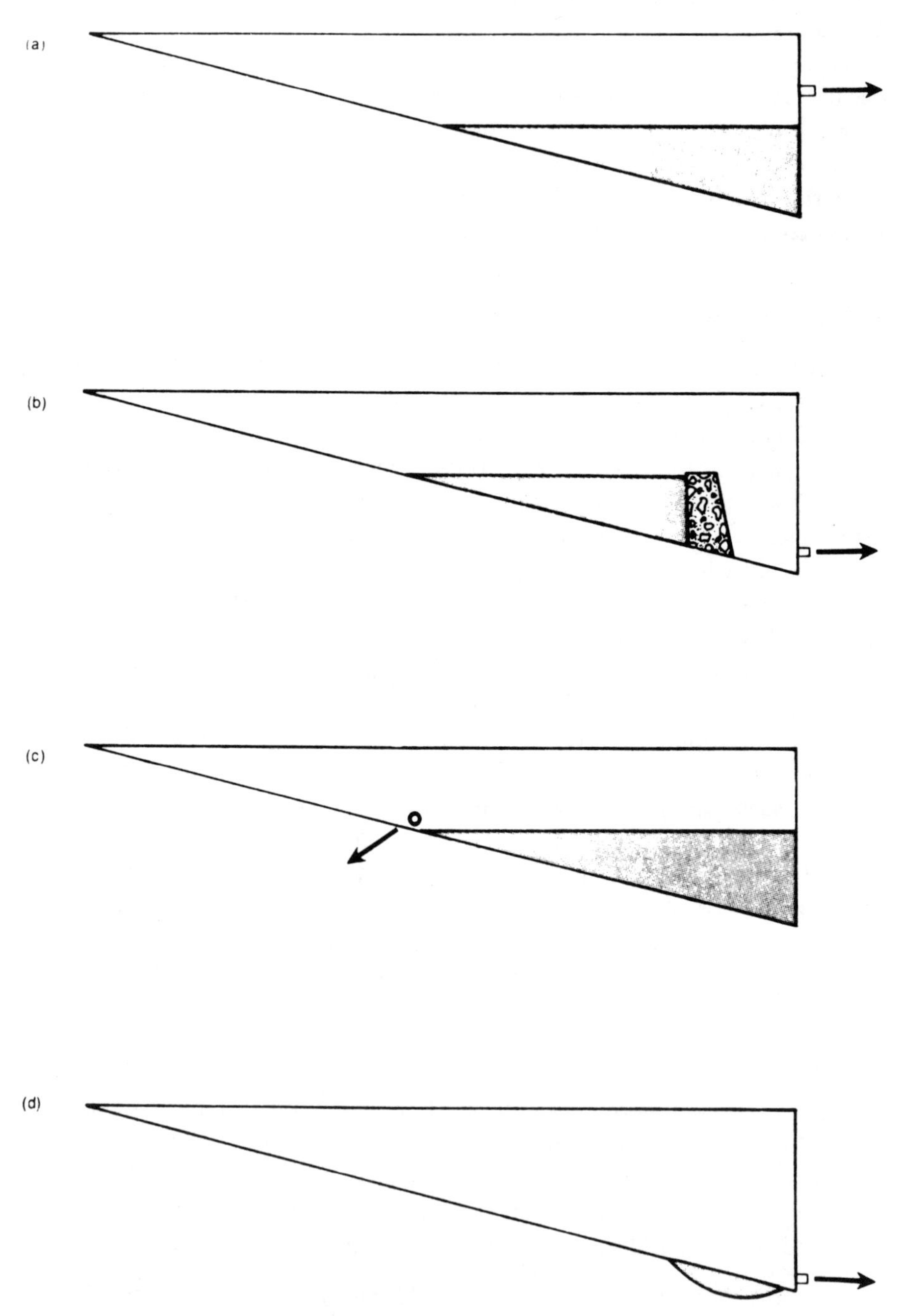

Figure 4.10 Akinetic spaces (shaded areas) due to (*a*) midlevel withdrawal current, (*b*) submerged weir, (*c*) outlet located upstream from the main dam, and (*d*) depression in the thalweg.

currents." The first situation that Hutchinson described occurred in Flaming Gorge Reservoir during the early years of impoundment (Wegner, personal communication). Soil salts going into solution from the inundated area and high-density inflows caused a monimolimnion to develop. In combination with midlevel withdrawal above the chemocline and temperature stratification, meromixis caused anoxic hypolimnetic conditions. After 18 years of operation a change to an epilimnetic discharge routed the low-density mixolimnetic water downstream, which reduced the density differences between the upper and lower layers. This resulted in an overturn of the reservoir the following spring.

The second situation that Hutchinson described has been observed in Keystone Reservoir (Eley 1967). Keystone Reservoir receives water from both the Cimarron River and the Arkansas River. The Cimarron River contained approximately four times the amount of total dissolved solids as the Arkansas River. As a result Cimarron water entered the reservoir as an underflow which, in combination with epilemnetic discharge, produced meromictic conditions. This resulted in an extreme clinograde oxygen curve in July (Figure 4.11*a*). An unusual negative heterograde DO curve was evident in January (Figure 4.11*b*). Colder, well-aerated water from the Cimarron River entered as an underflow that resulted in a negative heterograde curve in the monimolimnion. Hypolimnetic releases during the following year reduced chemical stratification, and complete overturn occurred in September. Continued hypolimnetic releases reduced chemical stratification, resulting in higher hypolimnetic DO concentrations in the summer of 1967 than in the previous summer (Figure 4.11*c*). It is interesting to note that in one reservoir a change from a hypolimnetic to an epilimnetic withdrawal resulted in overturn while in the other reservoir a change from an epilimnetic to a hypolimnetic withdrawal had the same result.

CONCLUSIONS AND IMPLICATIONS

An interaction of natural and regulated flow dynamics, temperature, morphology, and community metabolism results in a general pattern of hypolimnetic anoxic zone development following vernal stratification, which is common to deep-storage reservoirs. The hypolimnetic anoxic zone initially develops in the thalweg of the transition zone during summer stratification and then develops both upreservoir and downreservoir

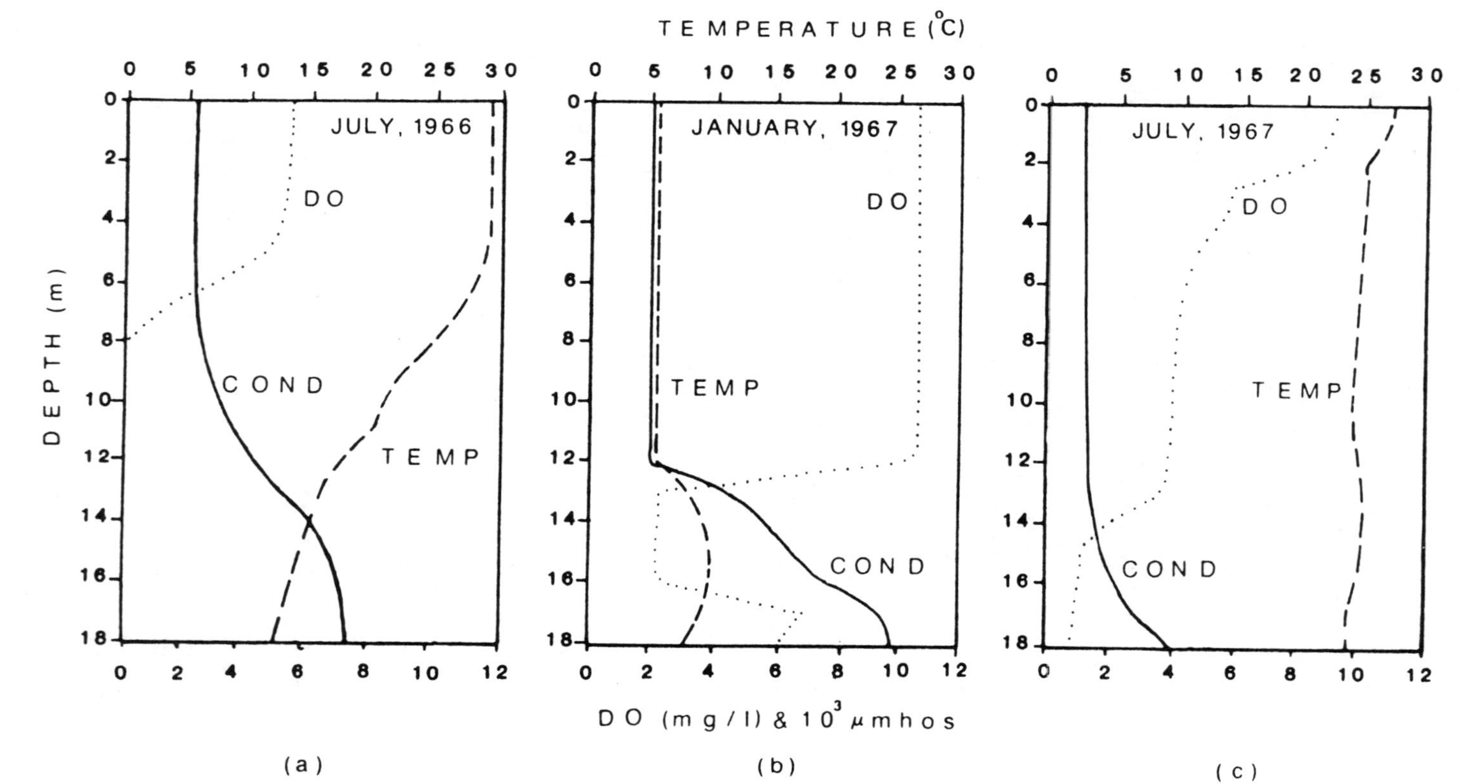

Figure 4.11 Distribution of temperature, conductivity, and dissolved oxygen in meromictic Keystone Reservoir (modified from Eley 1967). (*a*) Clinograde oxygen curve. (*b*) Negative heterograde oxygen curve caused by underflow of water with a high oxygen content. (*c*) Oxygen curve following overturn.

over time. Concurrently, the anoxic zone develops upward and laterally out of the thalweg. The time of onset and duration of hypolimnetic anoxia are primarily controlled by meteorologic conditions that vary annually according to geographic location.

This pattern of development has an important impact on sampling design, the interpretation of sampling results, and the overall dynamics of the ecosystem. Sampling design and interpretation of the results must take into account that the development of anoxic conditions in reservoirs may vary from year to year, depending upon the importance of various factors that control the location and development of the anoxic zone.

For example, during a year of low inflow-outflow the initial location and subsequent development of the anoxic zone will be shifted upreservoir. For a study involving parameters that are affected by anoxic conditions, upreservoir sampling will need to be more extensive than during a year of high inflow-outflow. The recycling of nutrients may also be more extensive during a year of low inflow-outflow in the upreservoir end and a subsequent increase in primary loading from the parent river. The extent of anoxic conditions must be known in order to correctly interpret results.

The development of the anoxic zone may have a significant impact on the overall dynamics of the ecosystem. During a year of high inflow-outflow the anoxic zone will reach the dam sooner than during a year of low inflow-outflow. For a bottom-draining reservoir this will result in a greater amount of nutrients being released downstream from the reservoir over the summer stratification period and, along with a decrease in the oxygen content of the water released downstream, may alter the makeup of the downstream ecosystem.

The most common variations in this general pattern of anoxic development include metalimnetic oxygen minima and maxima, oxygen blocks, akinetic spaces, and meromixis, all of which can effect the development of the hypolimnetic anoxic zone. They also have an effect on sampling design, interpretation of results, and the overall dynamics of the ecosystem. For example, a metalimnetic oxygen minimum may help recycle nutrients in the middle layers of a reservoir, promoting phytoplankton growth above the minimum if this area lies in the photic zone. Metalimnetic oxygen minima may also have an impact on the vertical migration of fish, the vertical distribution of bacteria, etc. Oxygen blocks may have a similar effect horizontally. In order to detect the occurrence of these variations and their effects on the ecosystem, sampling design must maximize the chances of their detection.

Although the general pattern of hypolimnetic anoxic zone development may vary depending upon whether the reservoir is a mainstream, transitional, or deep-storage reservoir, the development of the anoxic zone in a deep-storage reservoir exhibits a continuum of the development of the anoxic zone in these three reservoir types. A hypolimnetic anoxic zone may or may not develop in a mainstream-type reservoir. If it does it will form rapidly and may be quickly dissipated by large inflows or high winds. The same situation exists in the riverine zone of a deep-storage reservoir. Hypolimnetic anoxia in a transitional-type reservoir will initially develop close to the dam and upreservoir, depending upon wind and inflow. In a deep-storage reservoir the hypolimnetic anoxic zone initially develops in the transition zone, and the upreservoir development is also influenced by inflow and wind. In certain cases a deep-storage reservoir may be drawn down sufficiently to resemble a transitional- or mainstream-type reservoir. In this situation the DO dynamics would be expected to change accordingly.

REFERENCES

Baker, J. R., J. E. Deacon, T. A. Burke, S. S Egdorf, L. J. Paulson, and R. W. Tew. 1977. Limnological aspects of Lake Mead, Nevada-Arizona. U.S. Bur. Recl. Tech. Rep. No. REC-ERC-77-9.

Beadle, L. D. 1974. The inland waters of tropical Africa. Longman, Inc., New York., NY. 365 pp.

Bira, M. R. 1984. The effects of coves and metalimnetic oxygen minima on the development of hypolimnetic oxygen conditions in Canyon Lake (Reservoir), Texas. M. S. Thesis, Southwest Texas State University, San Marcos, TX. 50 pp.

Birge, E. A. and C. Juday. 1911. The inland lakes of Wisconsin. The dissolved gases and their biological significance. *Bull. Wis. Geol. Nat. Hist. Surv. 22*. 259 pp.

Bolke, E. L. 1979. Dissolved-oxygen depletion and other effects of storing water in Flaming Gorge Reservoir, Wyoming and Utah. Geological Survey Water Supply Paper 2058. 68 pp.

Bowmaker, A. P. 1976. The physico-chemical limnology of the Mwenda River Mountains, Lake Kariba. *Arch. Hydrolbiol.* 77:66–108.

Cangialosi, P. M. 1976. A phosphorus budget and lake models for Lake Ozonia. M. S. Thesis, Civ. Envir. Eng. Dept., Clarkson College of Technology, Potsdam, NY. 68 pp.

Charlton, M. N. 1980. Oxygen depletion in Lake Erie: Has there been any change? *Can. J. Fish. Aquat. Sci.* 37:72–81.

Churchill, M. A. 1958. Effects of impoundments on oxygen resources. Pages 107–130 in Proceedings of the seminar on the oxygen relationships of streams. R. A. Taft San. Eng. Ctr., Cincinnati, OH.

Churchill, M. A. and W. R. Nicholas. 1967. Effects of impoundments on water quality. *J. Sanit. Eng. Div. Proc. Am. Soc. Civ. Eng.* 93:73–90.

Dendy, J. A. 1945. Depth distribution of fish in relation to environmental factors, Norris Reservoir. *J. Tenn. Acad. Sci.* 20:114–131.

Dendy, J. A. and R. H. Stroud. 1949. The dominating influence of Fontana Reservoir on temperature and dissolved oxygen in the Little Tennessee River and its impoundments. *J. Tenn. Acad. Sci.* 24(1):41–51.

Drury, D. D. and R. A. Gearheart. 1975. Bacterial population dynamics and dissolved-oxygen minimum. *J. Am. Wat. Works Assoc.* 67:154–158.

Ebel, W. J. and C. H. Koski. 1968. Physical and chemical limnology of Brownlee Reservoir, 1962–1964. *Fish. Bull.* 67:295–335. Bureau of Commercial Fisheries, Biological Laboratory, Seattle, WA.

Edinger, J. E. and E. M. Buchak. 1977. A hydrodynamic, two-dimensional reservoir model: Development and test application to Sutton Reservoir, Elk River, West Virginia. Contract No. DACW 27-76-C-0089. U.S. Army Engineer Division, Ohio River, Cinncinati, OH.

Eley, R. L. 1967. Physicochemical limnology and community metabolism of Keystone Reservoir, OK. Ph.D. Thesis, Oklahoma State University, Stillwater, OK. 240 pp.

Ellis, M. M. 1940. Water conditions affecting aquatic life in Elephant Butte Reservoir. *Bull. U.S. Bur. Fish.* 49:257.

Falter, C. M. 1976. Early limnology of Dworshak Reservoir, North Idaho. Pages 285–294 in R. D. Andrews, R. L. Carr, F. Gibson, B. E. Land, R. A. Solter, and K. C. Swedburg, eds. Proceedings of the symposium on terrestrial and aquatic ecology studies of the Northwest. EWSC Press, Eastern Washington College, Cheney, WA.

Fiala, L. 1966. Akinetic spaces in water supply reservoir. *Verh. Internat. Verein. Limnol.* 16:685–692.

Fish. F. F. and R. A. Wagner. 1950. Oxygen block in the mainstream of the Willamette River. U.S. Fish Wildl. Serv. Spec. Sci. Rep. No. 41. 19 pp.

Frink, C. R. 1969. Chemical and mineralogical characteristics of eutrophic lake sediments. *Soil Sci. Soci. Am.* 33(3):369–372.

Gnilka, A. 1975. Some chemical and physical aspects of Center Hill Reservoir, Tennessee. *J. Tenn. Acad. Sci.* 50(1):7–10.

Goda, T. 1959. Density currents in an impounding reservoir. Int. Assoc. Hydr. Res., 8th Congr. Montreal, 3-C:1–29.

Gordan, J. A. 1980. An evaluation of the LARM two-dimensional model for water quality management purposes. Pages 518–527 in H. G. Stefan, ed. Proceedings of the symposium on surface water impoundments. Amer. Soc. Civ. Eng., New York, NY.

Gordan, J. A. and W. R. Nicholas. 1977. Effects of impoundments on water quality: Observations of several mechanisms of dissolved oxygen depletion. Div. Env. Planning, TVA, Chattanooga, TN. 55 pp.

Gordan, J. A. and B. A. Skelton. 1977. Reservoir metalimnion oxygen demands. *J. Sanit. Eng. Div. Proc. Am. Soc. Civ. Eng.* 103(EE6):1001.

Gordan, J. A. and J. W. Morris. 1979. Dissolved oxygen depletion mechanism operating in the metalimnion of a deep impoundment. Pages 29–38 in E. E. Driver and W. O. Wunderlich, eds. Environmental effects of hydraulic engineering works. T.V.A., Knoxville, TN.

Graneli, W. 1978. Sediment oxygen uptake in south Swedish lakes. *Oikos* 30:7–16.

Haberle, T. G. 1981. The spatial and temporal pattern of the depletion of hypolimnetic dissolved oxygen in Canyon Reservoir, Texas. M. S. Thesis, Southwest Texas State University, San Marcos, TX. 49 pp.

Hall, R. D. 1983. Patterns of dissolvled organic carbon and particulate organic carbon in the Guadalupe River and Canyon Lake (Reservoir). M. S. Thesis, Southwest Texas State University, San Marcos, TX. 57 pp.

Hannan, H. H. 1979. Chemical modifications in reservoir-regulated streams. Pages 75–94 in J. V. Ward and J. A. Stanford, eds. The ecology of regulated streams. Plenum Publishing Co., New York, NY.

Hoffman, D. A. and A. R. Jones. 1973. Lake Mead, a case study. Pages 220–223 in W. C. Ackerman, G. F. White, and E. B. Worthington, eds. Man-made lakes: Their problems and environmental effects. Am. Geophys. Union, Washington, DC.

Hrbacek, J., L. Prochazkova, V. Straskraboua-Prokesova, and C. O. Junge. 1966. The relationship between the chemical characteristics of the Vlstava River and Slapy Reservoir with an appendix: Chemical budget for Slapy Reservoir *Hydrobiol. Stud.* 1:41–84.

Hrbacek, J. L. and M. Straskraba. 1966. Horizontal and vertical distribution of temperature, oxygen, pH, and water movements in Slapy Reservoir. *Hydrobiol. Stud.* 1:7–40.

Hutchinson, G. E. 1957. A treatise on limnology. Vol. 1: Geography, physics and chemistry. John Wiley and Sons, Inc., New York, NY. 1015 pp.

Hyne, N. J. 1978. The distribution and source of organic matter in reservoir sediments. *Environ. Geol.* 2:279–287.

Iwanski, M. L., D. J. Bruggink, and J. W. Shipp. 1979. Field research on the effects of impoundments on water quality. Pages 395–406 in E. E. Driver and W. O. Wunderlich, eds. Environmental effects of hydraulic engineering works. T. V. A., Knoxville, TN.

Johnson, N. M. and F. W. Page. 1980. Oxygen depleted waters: Origin and distribution in Lake Powell, Utah-Arizona. Pages 1630–1637 in H. G. Stefan, ed. Proceedings of the symposium on surface water impoundments. Amer. Soc. Civ. Eng., New York, NY.

Krenkel, P. A., E. L. Thackston, and F. L. Porkov. 1968. The influence of impoundments on waste assimilative capacity. Pages 1–39 in R. A. Elder, P. D. Krenkel, and E. L. Thackston, eds. Proceedings of the specialty conference on current research into the effects of reservoirs on water quality. Tech. Rep. No. 17, Dept. Environ. Water Res. Eng.

Kusnetzov, S. I. and G. S. Karsinken. 1931. Direct method for the quantitative study of bacteria in water and some considerations on causes which produced a zone of oxygen minimum in Lake Glubokoje. *Zbl. Bakt., Sev. II* 83:169–174.

Larson, D. W. 1980. Limnology of selected reservoirs in the Oregon Cascade Range: Effects on water quality in the Willamette River. Pages 1526–1541 in H. G. Stefan, ed. Proceedings of the symposium on surface water impoundments. Amer. Soc. Civ. Eng., New York, NY.

Lasenby, D. C. 1975. Development of oxygen deficits in 14 Southern Ontario lakes. *Limnol. Oceanogr.* 20(6):993–999.

Latif, A. F. A. 1973. Effect of impoundment on Nile biota in Lake Naser. Pages 435–445 in W. W. Driver and W. D. Wunderlich, eds. Environmental effects of hydraulic engineering works. T. V. A., Knoxville, TN.

Lawrence, J. M. 1967. Dynamics of physicochemistry in a large reservoir. Pages 100–113 in Reservoir Fishery Resources Symposium. Am. Fish. Soc., Washington, DC.

Leentvaar, P. 1973. Lake Brokonpondo. Pages 186–196 in W. C. Ackerman, G. F. White, and E. B. Worthington, eds. Man-made lakes: Their problems and environmental effects. *Geophys. Mono. 17.* Amer. Geophys. Union, Washington, DC.

Lepak, C. J. 1976. Limiting nutrient and trophic level determination of Lake Ozonia by algal assay procedure. M. S. Thesis, Clarkson College of Technology, Potsdam, NY. 71 pp.

Louder, D. E. and W. D. Baker. 1966. Some interesting limnological aspects of Fontana Reservoir. Pages 1–16 in Proc. 20th Ann. Conf. SW Assoc. Game Fish Comm. Oct. 24–26, 1966. Asheville, NC.

Lund, J. W. G., F. J. H. Mackereth, and C. H. Mortimore. 1963. Changes in depth and time of certain chemical and physical conditions and of the standing crop of *Asterionella formosa* Has. in the north basin of Windemere in 1947. *Phil. Trans. Roy. Soc. Lond. Serv.* B246:255–290.

Lyman, F. E. 1944. Effects of a flood upon temperature and dissolved oxygen relationships in Cherokee Reservoir, Tennessee. *Ecology* 25:78–84.

Mullan, J. W., D. I. Morais, and R. L. Applegate. 1970. Thermal, oxygen and conductance characteristics of a new and an old reservoir. Tech. Paper No. 52, *Bur. Sport Fish. Wildl.* 68 pp.

Nix, J. 1974. Distribution of trace metals in a warm water release impoundment. Arkansas Water Resources Research Center, Fayetteville, AR. 337 pp.

Raheja, P. C. 1973. Lake Nasser. Pages 234–245 in W. C. Ackerman, F. F. White, and E. B. Worthington, eds. Man-made lakes: Their problems and environmental effects. *Geophy. Mono. 17.* Amer. Geophys. Union, Washington, DC.

Rawson, J. 1979. Water quality of Livingston Reservoir on the Trinity River, Southeastern Texas. TX Dept. Water Resour., Rep. 230. 79 pp.

Rettig, S. A. 1980. Limnological reconnaissance of Shasta Lake — Shasta County, California, March 1977–Sept. 1978. Pages 1474–1483 in H. G. Stefan, ed. Proceedings of the symposium on surface water impoundments. Amer. Soc. Civ. Eng., New York, NY.

Ruttner, F. 1963. Fundamentals of limnology, 3rd Ed. University of Toronto Press, Toronto, Canada. 295 pp.

Shapiro, J. 1960. The cause of a metalimnetic minimum of dissolved oxygen. *Limnol. Oceanogr.* 5:216–227.

Soltero, R. A, J. C. Wright, and A. A. Horpestad. 1974a. The physical limnology of Bighorn Lake — Yellowtail Dam Mountain: Internal density currents. *Northwest Sci.* 48(2):107–123.

Soltero, R. A., A. F. Gasperino, and W. G. Graham. 1974b. Chemical and physical characteristics of a eutrophic reservoir and its tributaries: Long Lake, Washington. *Wat. Res.* 8:419–431.

Soltero, R. A., A. F. Gaspenino, and W. G. Graham. 1975. Chemical and physical characteristics of a eutrophic reservoir and its tributaries: Long Lake, Washington — II. *Water Res.* 9:1059–1064.

Stroud, R. H. and R. G. Martin. 1973. Influence of reservoir discharge location on the water quality, biology, and sport fisheries of reservoirs and tail waters. Pages 540–548 in W. C. Ackerman, G. F. White, and E. B. Worthington, eds. Man-made lakes: Their problems and environmental effects. *Geophys. Mono. 17.* Am. Geophys. Union, Washington, DC.

Tenant, D. L., R. E. Thomas, and T. Gray. 1967. Physico-chemical limnology of reservoirs in Southeast Nebraska. Pages 537–540 in Reservoir fisheries resources symposium. Amer. Fish. Soc., Washington, DC.

Vanderhoof, R. A. 1965. Changes in waste assimilation capacity resulting from streamflow regulation. Pages 129–145 in Symposium on streamflow regulation for quality control. R. A. Taft Sanitary Eng. Ctr., Cincinnati, OH.

Weibe, A. H. 1941. Density currents in impounded waters — their significance from the standpoint of fisheries management. *Trans. 6th N. Am. Wildl. Conf.* 6:256–264.

Weibe, A. H. 1940. The effect of density currents upon vertical distribution of temperature and dissolved oxygen in Norris Reservoir. *J. Tenn. Acad. Sci.* 15:301–303.

Weibe, A. H. 1939a. Dissolved oxygen profiles at Norris Dam and in Big Creek sector of Norris Reservoir (1937), with a note on the oxygen demand of water (1938). *Ohio J. Sci.* 39:27–36.

Weibe, A. H. 1939b. Density currents in Norris Reservoir. *Ecology* 20:446–450.

Weibe, A. H. 1938. Limnological observations of Norris Reservoir with special reference to dissolved oxygen and temperature. *Trans. 4th N. Am. Wildl. Conf.* 4:440–457.

Wetzel, R. G. 1975. Limnology. W. B. Saunders Co., Philadelphia, PA. 767 pp.

Wiedenfeld, R. C. 1980. The limnology of Canyon Reservoir during years of contrasting flows. M. S. Thesis, Southwest Texas State University, San Marcos, TX. 58 pp.

Wunderlich, W. O. 1971. The dynamics of density-stratified reservoirs. Pages 219–231 in G. E. Hall, ed. Reservoir fisheries and limnology. Spec. Publ. 8, Amer. Fish. Soc., Washington, DC.

Young, W. C., H. H. Hannan, and J. W. Tatum. 1972. The physiochemical limnology of a stretch of the Guadalupe River, Texas, with five main-stream impoundments. *Hydrobiologia* 40(3):297–319.

CHAPTER 5

Reservoir Nutrient Dynamics

ROBERT H. KENNEDY AND WILLIAM W. WALKER

Limiting a discussion of nutrient dynamics to reservoirs risks the implication that there are inherent differences between nutrient-related processes occurring in reservoirs and those occurring in other types of aquatic systems. This is certainly not the intent of this chapter. However, there are a number of substantive differences between reservoirs, as a group, and most other lakes (Kennedy et al. 1985, Ryder 1978, Thornton et al. 1981), as has been discussed in preceding chapters in this volume.

The body of knowledge upon which our present understanding of nutrient dynamics is based has been acquired primarily from studies of small natural lakes. Since significant differences in flow regime and morphology do exist between reservoirs and small, natural lakes, it seems prudent to evaluate the relative importance of the various processes affecting nutrient distribution and availability in reservoirs. These processes include nutrient loading (external and internal), sedimentation, flow, mixing, and discharge. The intent of this chapter is to explore relations between these processes and nutrient conditions in reservoir ecosystems.

LOADINGS

Reservoir water quality and productivity are controlled to a large extent by the quantity and quality of external nutrient loadings. The nature of these nutrient inputs reflect, in turn, climatic regime and various watershed characteristics, including morphology, soil type, and land use. The strong regional variations in total phosphorus concentrations, reported by Omernik (1977) in his analysis of data from U.S. streams not impacted by point sources, reflect geographic variations in climatic and watershed characteristics. Streamflow, erosion rates, parent soil characteristics, and stream sediment transport properties are also regional in character and have potentially important effects on nutrient loadings and reservoir responses.

Canfield and Bachmann (1981) observed statistically significant differences between lakes and reservoirs in the parameters of an empirical model for predicting phosphorus sedimentation rate. They suggested that these differences might be attributed to "qualitative differences in the phosphorus inputs related to geographic location," specifically referring to the fact that most of the reservoirs included in their analysis were located in regions where large percentages of the phosphorus loadings were in particulate form.

The latitudes of 309 natural lakes in the United States sampled by the Environmental Protection Agency's (EPA) National Eutrophication Survey (U.S. Environmental Protection Agency 1978) are compared with the latitudes of 106 Corps of Engineers (CE) reservoirs also sampled by the EPA in Figure 5.1. The distribution of natural lakes is bimodal, with peaks in the North (glacial lakes) and the South (Florida lakes). Most of the CE reservoirs are located at intermediate latitudes (23–40 degrees N), where there are relatively few natural lakes. These latitudinal distributions are consistent with Canfield and Bachmann's suggestion, particularly since most of the CE reservoirs are located in unglaciated regions with high soil erodability and high watershed nutrient export.

WITHIN-LAKE PROCESSES

Flow

In reservoirs longitudinal gradients in physical, chemical, and biological factors result from the combined influence of hydrodynamics and basin morphology. While physical characteristics vary widely among reservoirs, reservoirs are often long and narrow and, unlike drainage lakes, receive water and nutrient inputs from a single, large tributary distant from the point of discharge. Although riverine effects dissipate with changes in basin width and depth along the axis of the reservoir, riverine influences

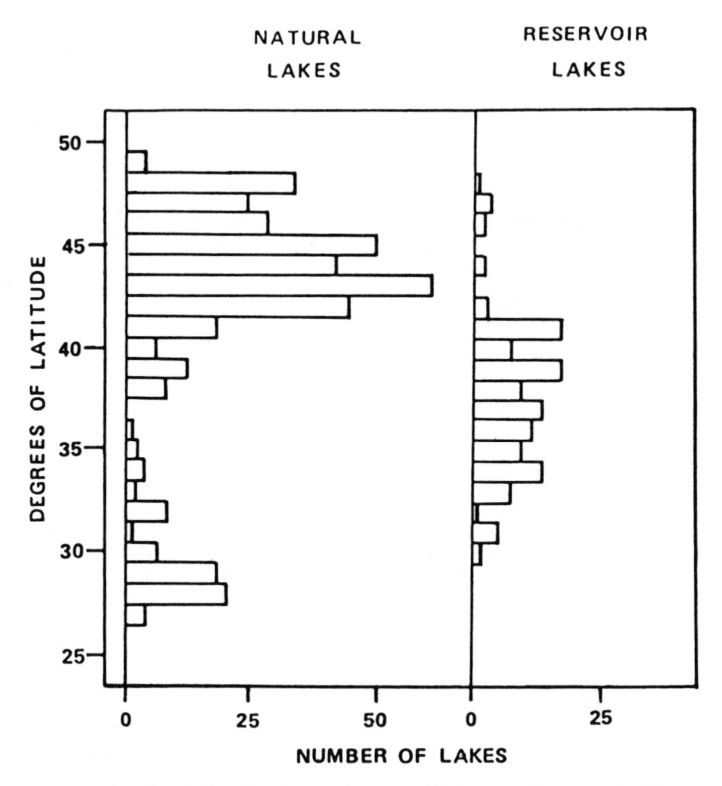

Figure 5.1 Latitudinal distributions of natural lakes and reservoir lakes sampled during EPA's National Eutrophication Survey (based on Walker 1981).

may often persist for great distances within some reservoirs. Thus, chemical and biological processes occur in a physical environment greatly influenced by flow regime, and as a result, processes acting to modify inflow water quality are ordered along a dynamic time/distance continuum governed by flow. This is in marked contrast to most natural lakes, in which vertical gradients resulting from thermal stratification predominate. Therefore, the relative importance of advective forces is one of the major distinctions between reservoirs and most other lakes.

The potential importance of interactions between hydrodynamic and morphometric characteristics and within-lake processes affecting nutrient-related attributes are most easily explored by considering idealized examples. For instance, consider the behavior of a nonconservative substance (e.g., phosphorus) in two nonstratified reservoirs of similar hydrology but dissimilar morphologies, one broad and deep and the other narrow and

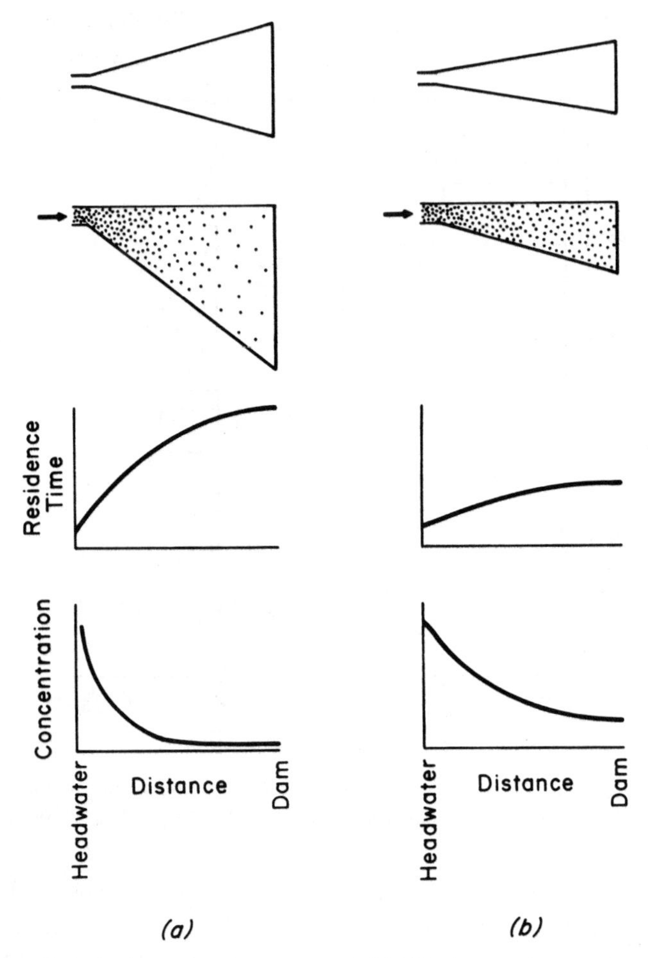

Figure 5.2 Distribution of inflow material (shading) and changes in water residence time and material concentration with distance from headwater to dam in two reservoirs receiving similar input but differing morphologic characteristics, one broad and deep (*a*) and the other narrow and shallow (*b*).

shallow (Figure 5.2). For the purpose of this discussion we assume that the nonconservative nature of the substance results from the occurrence of a first-order decay process. Since flows are similar, differences in water residence time are dictated by differences in basin morphology, with the broad, deep basin having a longer residence time. Therefore, the quantity of material retained in this basin would, based on residence time, be greater than that retained in the narrow, shallow basin. Of equal importance would be differences in concentration along the lake's length. In this example distance downstream from the river mouth is a surrogate measure of time. Therefore, concentrations along the length of the broad, deep lake would decline more sharply and to a greater extent than in the narrow, shallow basin.

This view of the establishment of gradients in nutrient concentration is, in general, consistent with observation (Peters 1979, Gloss et al. 1980, Thornton et al. 1981, Kennedy et al. 1982), suggesting that simple decay models may provide a means for describing concentration changes along gradients (Higgins and Kim 1981). Kennedy et al. (1982) were able to identify relics of previous hydrologic events along an otherwise predictable turbidity gradient in the reservoir West Point Lake (Atlanta, Georgia). These results suggest that longitudinal gradients in reservoirs reflect patterns resulting from both inflow history and reservoir operations.

The potential effects of inflow history are apparent if, in the above example, the inflow concentration is allowed to vary through time. Consider two nonstratified reservoirs with similar morphologies and inflows but dissimilar loading histories (Figure 5.3). In the reservoir receiving constant material input concentrations would, assuming a constant decay rate and a nondispersive flow regime, decline with distance along the length of the reservoir. However, concentrations along the length of the reservoir receiving inputs of variable concentration would exhibit a pattern reflecting loading history. Parcels of inflow water with low initial concentrations would follow and precede parcels of water with higher initial material concentrations. Although material concentrations for individual parcels of water would decline through time (and thus with distance), the decline in concentration along the length of the reservoir would be nonuniform.

The pattern becomes more complex under more realistic conditions of changing flow and concentration. In Figure 5.4 patterns in concentration observed at three points in time are more easily explained when viewed in a historical context. During baseflow, when inflow concentrations and flow rate are low, concentrations immediately below the headwater decline sharply and reach relatively low levels in downstream portions of the reservoir. Following a hydrologic event that increases both flow and inflow concentration, nutrient concentrations along the length of the reservoir increase as parcels of inflow water of higher initial concentration progress through the reservoir. Concentrations near the discharge,

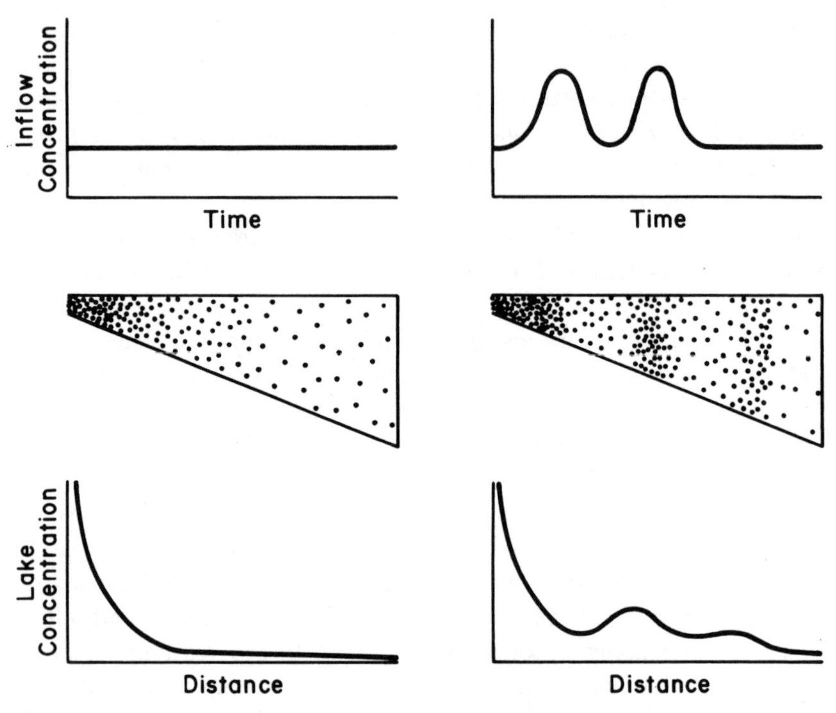

Figure 5.3 Material distributions along a reservoir receiving inflows with constant concentration (left) and a similar reservoir receiving inflows with variable concentration (right).

although still relatively low, would be higher than during baseflow. If reservoir conditions are again observed following the return of the tributary to baseflow conditions, remnants of earlier events would be recognized as areas of elevated concentration in the lower portion of the reservoir. Kennedy et al. (1982, 1981) documented such patterns in the reservoir Lake Red Rock (Iowa) following storm-related increases in nutrient and suspended solid loads from the Des Moines River. Although a general pattern of decreasing phosphorus concentration was observed from headwater to dam, elevated concentrations occurred at midlake coincident with the passage of the peak of the hydrograph.

Dispersion, assumed to be minimal in the preceding discussions, also influences the distribution of nutrients and the establishment of longitudinal gradients. As advective influences decrease with distance from the tributary inflow or because of increases in basin width and/or depth, wind-generated mixing becomes increasingly more important in distribut-

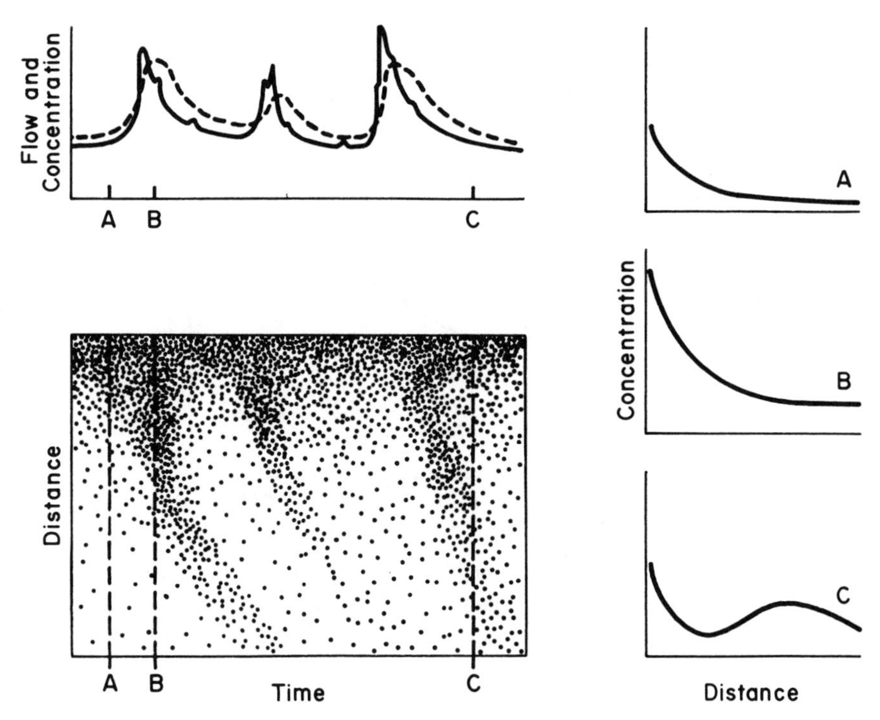

Figure 5.4 The combined effects of changes in tributary concentration and flow (upper left) on the spatial and temporal distribution of materials (shading) in the receiving reservoir (lower left). Longitudinal gradients in material concentration observed at three points in time (A, B, and C) are also depicted (right).

ing nutrients. In general, longitudinal gradients in nutrient concentrations are most pronounced in lakes in which advective or plug-flow conditions predominate, and gradients in lakes in which dispersion predominates are minimal. Walker (1982a) suggests a means by which reservoirs with a high potential for the development of longitudinal gradients can be distinguished from reservoirs that would be more completely mixed. Discriminatory factors include residence time, phosphorus sedimentation, and the relative importance of advection and dispersion as longitudinal transport processes (Figure 5.5). Maximal gradients would develop in reservoirs with long residence times, high sedimentation rates (i.e., high phosphorus retention), and an advectively dominated flow regime. Minimal longitudinal gradients would develop in reservoirs with short residence times, low sedimentation rates, and dispersive flow regimes.

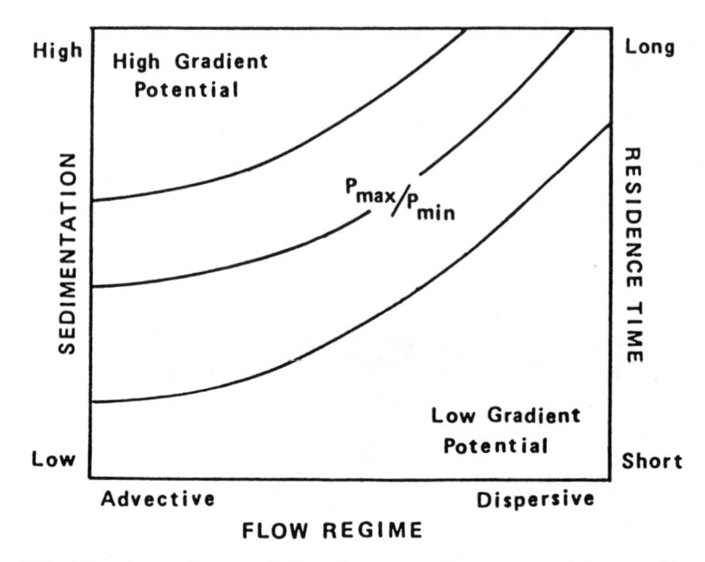

Figure 5.5 The dependence of phosphorus gradient potential on sedimentation rate, residence time, and flow regime. Lines of equal P_{max}/P_{min} (the ratio of maximum to minimum observed phosphorus concentration) are plotted. (Based on Walker 1982a.)

Density flows (see Chapter 2) also influence the distribution of nutrients in stratified reservoirs. During periods of thermal stratification vertical differences in density and the density difference between the inflowing river and reservoir water strata dictate the vertical placement of tributary inflows and thus the nutrients they transport. While riverine inflow patterns differ somewhat with changes in latitude, a general seasonal progression from overflows in fall, winter, and spring to interflows or underflows in summer is apparent (Wunderlich 1971, Carmack et al. 1979). While as yet incompletely evaluated these density-related phenomena have a significant impact on nutrient dynamics in stratified lakes and reservoirs.

If river inflows enter a lake, sink to a depth of comparable density, and then progress through the lake as an interflow at or below the thermocline, it seems reasonable to assume that the biological impact of riverborne ntutrients will be greatly diminished. During periods of interflow the epilimnion is at least partially isolated from an advected nutrient supply, and nutrient concentrations established prior to stratification would be expected to decline due to sedimentation losses. An increasingly limited supply of nutrients has obvious implications for summer phytoplankton production. Ultimately, the degree to which the epilimnion is isolated depends on the degree of density difference, the magnitude of flow, and basin morphology.

Field observations suggest that while nutrient loading to epilimnia may be greatly reduced during periods of interflow, mixing of riverine and epilimnetic water does occur. Kennedy et al. (1982) documented the movement of Chattahoochee River water through the upstream reach of West Point Lake using fluorescent dye as a water-mass tracer. During a nonstratified period river water mixed vertically and progressed through the reservoir as a plug flow. When the reservoir was stratified inflows progressed to the plunge point as a well-mixed plug flow and then were confined to a zone near the thermocline. While maximum dye concentrations were observed near the thermocline, significant quantities of dye were also found in surface waters, indicating the mixing of river water into the epilimnion. Kennedy et al. (1982) also observed the entrainment of nutrient-rich hypolimnetic water by the riverine interflow. Similar observations were made in Slapy Reservoir by Hrbacek et al. (1966). Thus, mixing across turbulent interfaces between the riverine layer and the epilimnion or between the shallow upstream portion of the hypolimnion and the riverine layer provides a mechanism by which nutrients may be redistributed vertically (Figure 5.6).

Carmack and Gray (1982) suggest another means by which nutrients transported by an interflowing riverine layer may enter the epilimnion during stratified periods. In Kootenay Lake, a long, deep, natural lake of the Columbia River System, nutrient-laden waters from the Kootenai and Duncan Rivers enter from opposite ends of the lake and, during summer months, are confined to a riverine-layer interflow located immediately above the thermocline. Exchanges between the riverine layer and the euphotic zone occur only following episodic mixing events. Internal seiche motion displaces the riverine layer upward where, under the influence of wind-generated mixing, water and nutrients of the riverine layer are entrained into the euphotic zone. Thus, nutrient concentrations in the euphotic zone, established at the onset of thermal stratification, decline during the summer because of a lack of nutrient inputs from the

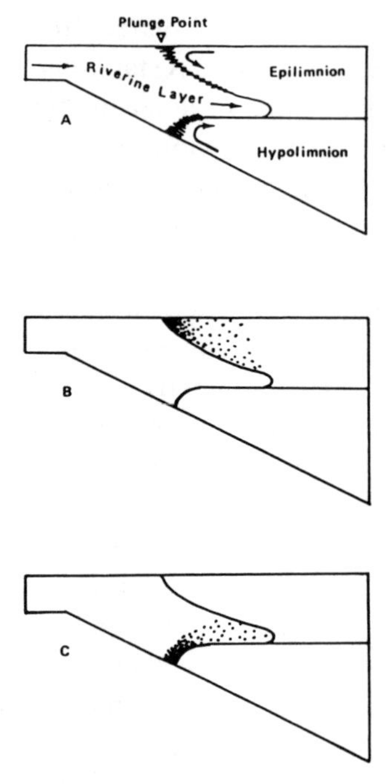

Figure 5.6 Turbulent interfaces (zigzag line) between epilimnion, hypolimnion, and an interflowing riverine layer (*a*). Exchanges of material from riverine layer to epilimnion (*b*) and from hypolimnion to riverine layer (*c*) are also indicated (shading).

river until such time that conditions favor entrainment of riverine water. These episodic events then augment dwindling epilimnetic nutrient supplies (Figure 5.7).

Sedimentation

Sedimentation and subsequent sediment-water interactions are major regulatory processes influencing the nutrient status of lakes and reservoirs. Sedimentation of nutrients, which can occur by the settling of particles or by the association of dissolved substances with settling biotic and abiotic particles, results in a loss of nutrients from the water column. Once deposited, sedimented materials may buffer or otherwise modify nutrient concentrations in overlying water. Both regulatory processes are influence by flow regime, reservoir morphology, tributary loading, trophic state, and the presence of gradients.

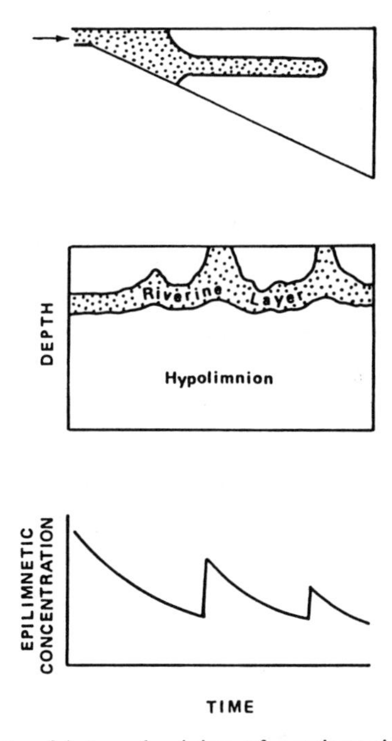

Figure 5.7 The effects of internal mixing of nutrient-rich riverine water (shading) and epilimnetic water on epilimnetic nutrient concentrations (based on Carmack and Gray 1982).

Differences in the relative contributions of particulate loads to reservoirs are reflected in higher sedimentation rates for reservoirs than for natural lakes. Canfield and Bachmann (1981), in comparing over 700 lakes and reservoirs, suggest that differences in phosphorus sedimentation may be due to the association of phosphorus with readily settleable allochthonous particulates. Higgins and Kim (1981) found similar differences when applying empirical loading-response models to Tennessee Valley Authority reservoirs. The apparent settling velocity of phosphorus for these lakes was nine times that for other types of lakes. They also identified differences between reservoir types. Tributary storage reservoirs, which are deeper and have longer residence times, retained a higher percentage of inflowing phosphorus than did shallower, more rapidly flushed reservoirs.

Sedimentation, like other nutrient-related processes in flow-dominated lakes and reservoirs, exhibits longitudinal gradients, with the highest sedimentation rates generally occurring in the portion of the nearest inflow. A majority of the material loads to Lake Red Rock, a short-retention-time reservoir on the sediment-laden Des Moines River, is retained in headwater areas near the river inflow (Kennedy et al. 1981). Significant losses (>75%) of inflow phosphorus occurring coincident with sedimentation of suspended solids resulted in relatively low concentrations in downstream areas, even during storm events. Sedimentary losses of phosphorus, nitrogen, and carbon are of greatest significance in the immediate vicinity of the inflow of the Caddo River to DeGray Lake, a long-retention-time storage improvement in south-central Arkansas. Sediment trap data reported by James et al. (1987) indicate that the greatest losses of allochthonous and autochthonous nutrients, metals, and organic matter occurred uplake near the river mouth. Differences between sedimentation rates for vertically arranged sets of traps also suggest higher sedimentation below the riverine layer during periods of interflow. Sedimentation losses are the prime cause for observed nutrient gradients in this lake (Thornton et al. 1981).

Seasonal trends in sedimentation are also observed. Nutrient-containing material collected by sediment traps deployed during spring in Lake Memphremagog was primarily inorganic, suggesting allochthonous inputs as the primary source (Spiller 1977). During summer and fall losses of nutrients were associated with sedimentary losses of autochthonous organic material. Similar seasonal trends were concluded from sediment trap data from DeGray Lake (James et al. 1987). Sedimentation patterns may also be influenced by seasonal changes in flow regime. In Lake Mojave cold, upstream releases from Lake Mead progress through the lake in summer as underflow, intercepting sedimenting autochthonous matter and transporting it downlake. Subsequent decomposition results in marked hypolimnetic gradients in inorganic nitrogen concentrations (Priscu, Verduin, and Deacon 1981). High nitrogen concentrations in Hoover Dam hypolimnetic releases from Lake Mead may also account for nitrogen gradients in Lake Mojave.

Patterns in sediment quality, important for understanding exchanges between bottom sediments and overlying water, result from interactions between hydrodynamic, morphometric, loading, and trophic state characteristics (Gunkel et al. 1984). Here again, longitudinal gradients are apparent. Hyne (1978) contrasts the distribution of organic matter in two midwestern reservoirs: Fort Gibson Lake, which receives low sediment inputs, and Lake Texoma, a productive reservoir receiving high sediment inputs. In Fort Gibson Lake organic materials are deposited in deep downlake areas. Deposits in Lake Texoma occur in shallow upstream areas as well as in deep areas of the lake, reflecting the combined influences of allochthonous organic matter inputs from the Red and Washita Rivers and autochthonous production.

Sediment characteristics in Lake Mulwala, an impoundment of the Murray River, Australia, are related to flow characteristics (Hart et al. 1976). Sediments underlying riverine areas of the lake are characteristically low in organic matter and total phosphorus. Sediments in areas little affected by river flow are high in organic matter and phosphorus, indicating the potential importance of flow and nutrient particle associations in the sorting of deposited materials. This result is supported by a comparative evaluation of the sediment characteristics of four CE reservoirs of varying trophic state and flow regime (Gunkel et al. 1984). In general, sediments along riverine reaches of these lakes were high in inorganic carbon but low in nutrients, metals, and organic carbon. In West Point Lake, the largest of the four reservoirs examined, sediment nutrient concentrations were highest downlake from the area of maximum phytoplankton production and the area influenced by riverine inflows. Coincident with increases in nutrient and organic content were increases in sediment moisture content.

Hakanson (1977) compared data for sediment samples collected from various locations in Lake Vanern and concluded that the nature of sedimentary environments can be deduced from sediment moisture content. Sediments in flow-dominated or high-energy areas (e.g., near inflows or areas of mixing) had low moisture content and low organic matter concentrations, indicating areas of sediment erosion and transport. Sediments in deeper areas or in areas less affected by turbulence had high moisture content and high organic matter concentrations, indicating that these are areas of sediment accumulation. Adopting similar criteria, Gunkel et al. (1984) were able to differentiate between sediments from advectively dominated areas of reservoirs and those from areas where nutrient accumulation in sediments occurs.

Relations between nutrient loading, sedimentation, flow, productivity, and sediment quality can be suggested (Figure 5.8). Inflow nutrient concentrations decline along a gradient from headwater to dam due to sedimentary losses. The losses result initially from the reduced carrying capacity for suspended particulates in uplake portions of the reservoirs, but later and further downlake because of phytoplankton uptake and settling. Phytoplankton production may be low in the headwaters due to

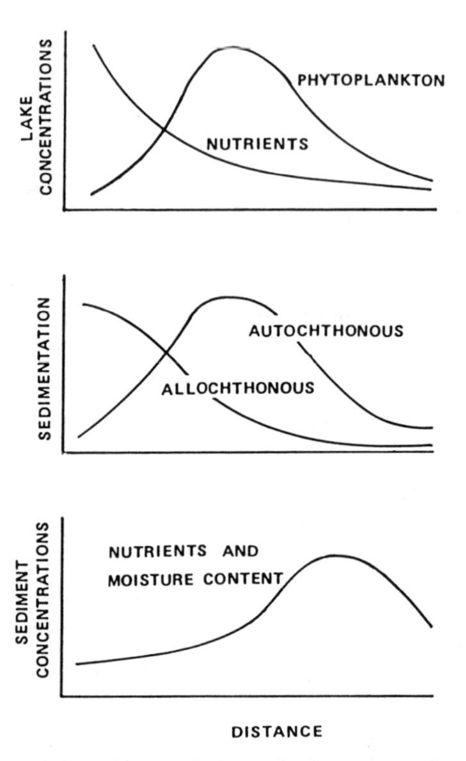

Figure 5.8 Changes in nutrient and phytoplankton (upper), autochthonous and allochthonous sedimentation rates (middle), and sediment and nutrient content (lower) with distance from headwater to dam.

inorganic turbidity and flushing but often increases downlake. Coincident with headwater declines in nutrient concentrations and mid- or downlake increases in phytoplankton production would be peaks in allochthonous and autochthonous sedimentation rates, respectively. Settling organic detritus and small inorganic particulates (e.g., clays and fine silts), both of which are often enriched with nutrients, are displaced downlake by advection. The net result of these processes is the accumulation of organic, nutrient-rich sediments of high moisture content in deeper downlake areas.

Internal loading

Seasonal releases of nutrients from storage sites within a lake or reservoir (e.g., from the sediments) can have a pronounced impact on lake nutrient status, particularly during periods when inputs from external sources are minimal (Cooke et al. 1977). Despite the dominance of river-borne nutrient inputs internal nutrient recycling is also of ecological significance in reservoirs. And, considering the establishment of gradients in sediment characteristics, large differences in the degree of ecological significance may occur between locations within a particular reservoir. Occoquan Reservoir in northern Virginia, for instance, exhibits longitudinal differences in sediment phosphorus concentrations and differences in the degree of exchange between sediments and the overlying water (To and Randall 1975). Surficial uplake sediments in this reservoir have higher phosphorus concentrations and release rates than do sediments at downlake locations. Differences are also apparent for release/absorption threshold concentrations. Phosphorus releases from the anoxic uplake sediments ceased at water column concentrations above 1.75 mg P/l while downlake sediments sorbed phosphorus from bottom water when concentrations exceeded only 0.5 mg P/l.

Thus, internal loading from sediments is related to trophic state, oxygen dynamics, and sedimentary history. DeGray Lake, because of a history of variable conditions, provides an instructive example of these relationships (Kennedy and Nix 1987). Timber, forest litter, and other organic detritus in the reservoir basin were left relatively undisturbed prior to impoundment, and exerted a significant demand on dissolved oxygen supplies. Anoxic conditions developed immediately following the onset of stratification, and the entire hypolimnion exhibited anoxia during summer. Hypolimnetic nutrient concentrations, despite relatively low external loads, were high during this early period, suggesting the potential for significant exchanges of dissolved materials across the thermocline by mixing diffusion. In subsequent years, as pre-impoundment organic deposits were exhausted, oxygen conditions improved. Summer increases in hypolimnetic nutrient concentrations also diminished. Conditions are now more consistent with the lake's low loading rate and moderate productivity. However, internal loading still occurs in the headwater areas of the lake (Kennedy et al. 1986). Particulate organic loads from the Caddo River, which supplies a majority of the lake's nutrient and organic loads, are deposited on headwater sediments, resulting in the establishment of anoxic conditions during summer months. Coincident increases in hypolimnetic nutrient concentrations also occur. Sediment trap data suggest that exchanges of nutrients from this shallow headwater portion of the hypolimnion to the epilimnion may influence the establishment of maximal chlorophyll standing crops in the lake's headwater area.

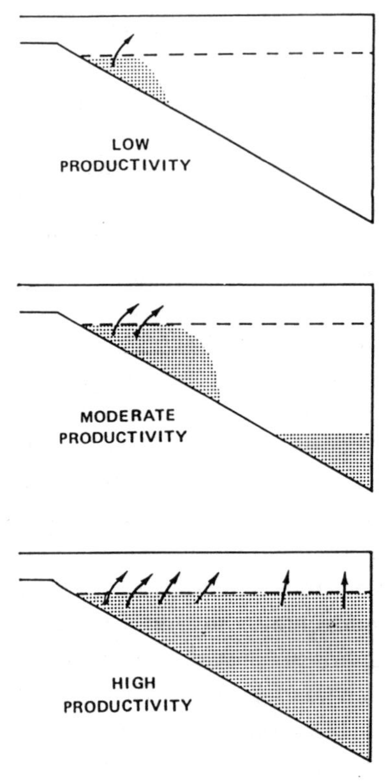

Figure 5.9 Longitudinal occurrences of hypolimnetic anoxia and nutrient increases (shading), and upward exchanges of nutrients (arrows) in reservoirs of differing productivity and loading.

Generalizations concerning potential longitudinal differences in the occurrence and significance of internal loading in reservoirs of similar hydrology and morphology, and its relation to reservoir trophic conditions, are presented in Figure 5.9. In unproductive reservoirs low allochthonous organic inputs and low autochthonous production may lead to anoxic conditions of limited extent (see Chapter 4). Releases from anoxic headwater sediments and turbulent mixing would lead to the possible introduction of nutrients to surface waters. In highly productive stratified reservoirs inputs of both allochthonous and autochthonous organic matter can result in a completely anoxic hypolimnion and consequent increases in nutrient releases. Concentration gradients across the thermocline could result in the introduction of nutrients to the epilimnion at downlake locations. Internal loadings of intermediate significance would occur in reservoirs of moderate productivity.

RESERVOIR OPERATION

The manner in which water is discharged differs in natural lakes and reservoirs. Differences are also apparent in the degree of daily or seasonal fluctuation in water level. Natural lakes discharge water at the surface and, since discharges are uncontrolled, lake level changes are seldom extreme. Reservoirs, on the other hand, are designed to modify or control river flows and therefore often experience significant changes in water level. In hydropower reservoirs levels can fluctuate dramatically over diel cycles. While surface releases do occur, subsurface discharges are more common for large reservoirs. Reservoir operation might, therefore, be expected to influence nutrient dynamics.

Wright (1967) hypothesized that surface-discharge lakes trap nutrients and dissipate heat while subsurface-discharge lakes dissipate nutrients and store heat. Martin and Stroud (1973) and Martin and Arneson (1978) provided supporting evidence by comparing reservoirs of differing operation, and a reservoir and natural lake, respectively. Total organic carbon and total phosphorus concentrations were proportionately higher in the tailwater of Nolin Lake, a hypolimnetic-discharge reservoir, than in the Barren River Reservoir, which discharges epilimnetic water. Similar differences were demonstrated for Heglen Lake, a deep-discharge reservoir, and Quake Lake, a surface-discharge landslide lake.

Many nonhydropower reservoirs operate during summer months (when flows are seasonally low) as epilimnetic- or surface-discharge lakes. However, when flood waters are discharged, releases must be made using large hypolimnetic floodgates. The impact of such an operational change can often be dramatic. The use of hypolimnetic floodgates at Lake Red Rock following a summer storm event influenced flow patterns in the reservoir and resulted in the "short-circuiting" of inflowing nutrient-laden storm water. The actual residence time of storm water in the reservoir was half the theoretical residence time estimated from flow and lake volume data (Kennedy et al. 1981).

If nutrients accumulate in the reservoir hypolimnion, then hypolimnetic discharges, either by the release of flood waters or during hydropower generation, result in nutrient losses from the reservoir. The management implications of such losses are worthy of consideration. If the residence times of various depth strata can be manipulated by the operation of the outlet structure (e.g., by selective withdrawal), then reservoir managers could exert some control over nutrient-related processes occurring in reservoirs. For instance, nutrients transported by the riverine layer in stratified reservoirs could be "drained off" by withdrawing water from the depth of interflow. Simiarly, nutrients that enter reservoirs following storm events or those that accumulate in deep water during stratification could be selectively flushed from the reservoir, thus minimizing their impacts on algal production, as shown by Elser and Kimmel (1985). However, nutrient releases from one reservoir may constitute nutrient inputs to reservoirs located downstream and thus influence the productivity of these downstream systems.

Lake level changes also influence nutrient conditions. In addition to changes in water residence time, fluctuating water levels can increase nutrient exchanges between littoral and pelagic zones. Metabolic activities in the macrophyte-infested littoral zone of Kremenchug Reservoir increased nitrogen and phosphorus concentrations during periods of constant lake level (Zimbalevskaya et al. 1976). During hydropower releases lake level dropped and nutrients were drawn from the littoral to the pelagic zone, where they were available for phytoplankton. The return of Quabbin Reservoir to a normal level after a six-year drought resulted in the inundation of newly established terrestrial vegetation around the perimeter of the lake. Miner (1974) calculated nitrogen and phosphorus loads from the decoposition of this material to be 1.3 g N/m^2 and 0.1 g P/m^2, respectively. Nutrient exchanges of this type are probably of greatest significance in reservoirs with high shoreline development ratios.

GENERALIZATIONS

Reservoirs, like all lakes, are dynamic aquatic ecosystems of which we have only a limited understanding. However, based on the preceding discussions we can offer several generalizations about nutrient dynamics in reservoirs. The organization of these ideas is aided by the suggestion that reservoirs can be divided into riverine, transition, and lacustrine zones, but it must be emphasized that the boundaries between zones are often difficult to delineate and that the locations of zones are temporally unstable.

An advective flow regime in combination with a long, narrow basin morphology results in the spatial ordering of nutrient-related processes and the establishment of gradients from headwater to dam. Nutrient loads, a large percentage of which are often associated with suspended particulates, progress through the riverine zone and are deposited downlake as flows diminish. Nutrient availability for phytoplankton production is reduced by sedimentation losses and in stratified water columns by the downward displacement of density flows in the transition zone. Nutrient utilization by phytoplankton is potentially greatest in this zone near the boundary between the transition and riverine zones. Nutrient availability is further diminished in the lacustrine zone, and vertical exchanges may provide important nutrient supplies for phytoplankton growth. The spatial distribution of nutrients, which is also affected by outlet operation, may in turn be reflected by patterns in sediment quality.

IMPLICATIONS FOR EMPIRICAL MODELING

Nutrient loading models were originally developed from northern lake data and designed to predict the eutrophication of a water body as a function of three primary variables:

P_i = annual inflow total phosphorus concentration (mg/m^3)
Z = mean depth (m)
T = annual mean hydraulic residence time (years)

Most published models can be expressed in combinations of the above terms (Vollenweider 1968, 1976, Chapra 1975, Larsen and Mercier 1976, Jones and Bachmann 1976, Canfield and Bachmann 1981).

The models are empirical in nature and should not be used outside the range of the data set used for model calibration. The application "range" refers not only to the three explicit variables mentioned but also to other characteristics (including impoundment type and region) that may influence nutrient dynamics and are therefore implicit in the model formulation.

Lake-reservoir differences of potential significance to the development and use of loading models include:

1. Reservoirs tend to have shorter hydraulic residence times, which may indicate the need to formulate nutrient and water balances on seasonal, as opposed to annual, bases.
2. Because of regional geographic factors, reservoirs tend to have higher percentages of particulate phosphorus loadings and higher

sediment accumulation rates, both of which may influence the parameter estimates of phosphorus retention models.

3. Reservoirs tend to have greater concentrations of allochthonous suspended solids, which influence phosphorus-chlorophyll and chlorophyll-transparency relationships.
4. Reservoir morphometric and hydrodynamic characteristics are more conducive to the development of spatial gradients in phosphorus and related trophic-state indicators; predictions of spatially averaged conditions may not adequately describe many reservoirs.
5. Hydrodynamic factors (underflows, interflows, bottom outlets, fluctuations in pool level) influence nutrient responses. These factors are not directly accounted for in existing model formulations.

These differences, which may have important effects on model formulations, parameter estimates, monitoring, and data-reduction procedures, must be considered if appropriate reservoir eutrophication assessments are to be developed.

Empirical studies by Clausen (1980), Canfield and Bachmann (1981), Higgins and Kim (1981), Walker (1982b, 1982c, 1985), and Mueller (1982) have shown that the parameter estimates of phosphorus retention models originally developed from northern lake data require significant adjustments when the models are recalibrated to reservoir data sets. These adjustments generally reflect higher phosphorus sedimentation rates in reservoirs as compared with natural lakes for a given inflow concentration, mean depth, and hydraulic residence time.

REFERENCES

Canfield, D. E. and R. W. Bachmann. 1981. Prediction of total phosphorus concentrations, chlorophyll-a and Secchi disc in natural and artificial lakes. *Can. J. Fish. and Aq. Sci.* 38:414–423.

Carmack, E. C., C. B. J. Gray, C. H. Pharo, and R. J. Daley. 1979. Importance of lake-river interactions on seasonal patterns in the general circulation of Kamloops Lake, British Columbia. *Limnol. Oceanogr.* 24:634–644.

Carmack, E. C. and C. B. J. Gray. 1982. Patterns of circulation and nutrient supply in a medium residence time reservoir Kootenay Lake, British Columbia. *Can. Wat. Res. J.* 7:51–70.

Chapra, S. C. 1975. Comment on "An empirical method of estimating the retention of phosphorus in lakes" by W. B. Kirchner and P. J. Dillon. *Wat. Resourc. Res.* 11:1033–1034.

Clausen, J. 1980. OECD Cooperative Programme for Monitoring Inland Waters — Regional Project — Shallow Lakes and Reservoirs. Organization for Economic Cooperation and Development.

Cooke, G. D., M. R. McComas, D. W. Waller, and R. H. Kennedy. 1977. The occurrence of internal phosphorus loading in two small, eutrophic, glacial lakes in northeastern Ohio. *Hydrobiol.* 56:2, 129–135.

Elser, J. J. and B. L. Kimmel. 1985. Nutrient availability for phytoplankton production in a multiple-impoundment series. *Can. J. Fish. Aquat. Sci.* 42: 1359–1370.

Gloss, S. P., D. E. Kidd, and L. M. Mayer. 1980. Advective control of nutrient dynamics in the epilimnion of a large reservoir. *Limnol. Oceanogr.* 24:219–228.

Gunkel, R. C., R. F. Gaugush, R. H. Kennedy, G. E. Saul, J. H. Carroll, and J. Gauthey. 1984. A comparative study of sediment quality in four reservoirs. Technical Report E-84-2. U.S. Army Engineer Waterways Experiment Station, Vicksburg, MS.

Hakanson, L. 1977. The influence of wind, fetch, and water depth on the distribution of sediments in Lake Vanern, Sweden. *Can. J. Earth Sci.* 14:397–412.

Hart, B. T., R. J. Mcgregor, and W. S. Perriman. 1976. Nutrient status of the sediments in lake mulwala. *I. Total Phosphorus. Aust. J. Mar. Freshwat. Res.* 27:129–135.

Higgins, J. M. and B. R. Kim. 1981. Phosphorus retention models for Tennessee Valley Authority reservoirs. *Wat. Resourc. Res.* 17:571–576.

Hrbacek, J., L. Prochazkova, V. Straskraboua-Prokesova, and C. O. Junge. 1966. The relationship between the chemical characteristics of the Vlstara River and Slapy Reservoir with an appendix: Chemical budget for Slapy Reservoir. *Hydrobiol. Stud.* 1:41–84.

Hyne, N. J. 1978. The distribution and source of organic matter in reservoir sediments. *Env. Geol.* 2:279–287.

James, W. F., R. H. Kennedy, R. H. Montgomery, and J. Nix. 1987. Seasonal and longitudinal variations in apparent deposition rates within an Arkansas Reservoir. *Limnol. Oceanogr.* 32:5, 1169–1176.

Jones, J. R. and R. W. Backmann. 1976. Prediction of phosphorus and chlorophyll levels in lakes. *J. Wat. Poll. Contr. Fed.* 48:2176–2182.

Kennedy, R. H., K. W. Thornton, and J. H. Carroll. 1981. Suspended sediment gradients in Lake Red Rock. Pages 1318–1328 in H. G. Stefan, ed. Proceedings of the symposium on surface water impoundments. Amer. Soc. Civil Engr., New York, NY.

Kennedy, R. H., K. W. Thornton, and R. C. Gunkel, Jr. 1982. The establishment of water quality gradients in reservoirs. *Can. Wat. Res. J.* 7:71–87.

Kennedy, R. H., R. H., K. W. Thornton, and D. Ford. 1985. Characterization of the reservoir ecosystem. In D. Gunnison, ed. Microbial processes in reservoirs. Dr. W. Junk Publishers, Boston, MA.

Kennedy, R. H., R. H., W. F. James, R. H. Montgomery, and J. Nix. 1986. The influence of sediments on the nutrient status of DeGray Lake, Arkansas. In P. G. Sly, ed. Sediments and water interactions. Springer-Verlag Publishers, New York, NY.

Kennedy, R. H., R. H., and J. Nix, eds. 1987. Proceedings of the DeGray Lake Symposium. Technical Report E-87-4. US Army Engineer Waterways Experiment Station, Vicksburg, MS.

Larsen, D. P. and H. T. Mercier. 1976. Phosphorus retention capacity of lakes. *J. Fish. Res. Bd. Con.* 33:1742–1750.

Martin, R. G. and R. H. Stroud. 1973. Influence of reservoir discharge location on water quality, biology and sport fisheries of reservoirs and tailwaters. 1968–1971. U.S. Army Engineer Waterways Experiment Station, Vicksburg, MS.

Martin, R. G., D. B. and R. D. Arneson. 1978. Comparative limnology of a deep-discharge reservoir and a surface-discharge lake on the Madison River, Montana. *Freshwat. Biology*. 8:33–42.

Miner, N. H. 1974. The potential for impact for inundation of terrestrial vegetation on the water quality of Quabbin Reservoir — Commonwealth of Massachusetts. *Wat. Res. Bul.* 10(6):1288–1297.

Mueller, D. K. 1982. Mass balance model estimation of phosphorus concentrations in reservoirs. *Wat. Res. Bull.* 18:377–382.

Omernik, J. M. 1977. Nonpoint source-stream nutrient level relationships: A nationwide study. Corvallis Environmental Research Laboratory. EPA-600/3-77-105, U.S. Environmental Protection Agency.

Peters, R. H. 1979. Concentration and kinetics of phosphorus fractions along the trophic gradient of Lake Memphremagog. *J. Fish. Res. Board Can.* 36:970–979.

Priscu, J. C., J. Verduin, and J. E. Deacon. 1981. The fate of biogenic suspersoils in a desert reservoir. Pages 1657–1667 in H. G. Stefan, ed. Proceedings of the symposium on surface water impoundments. Amer. Soc. Civil Engr., New York, NY.

Ryder, R. A. 1978. Ecological heterogenity between north-temperate reservoirs and glacial lakes systems due to differing succession rates and cultural uses. *Verh. Int. Verein. Limnol.* 20:1568–1574.

Spiller, G. B. 1977. A mathematical model of seasonal and spatial variation in phosphorus concentrations in the surface waters of Lake Memphremagog, Quebec. M.S. Thesis, Biology Dept., McGill University, Montrea, Quebec.

Thornton, K. W., R. H. Kennedy, J. H. Carroll, W. W. Walker, R. C. Gunkel, and S. Ashby. 1981. Reservoir sedimentation and water quality — an heuristic model. Pages 654–661 in H. G. Stefan, ed. Proceedings of the symposium on surface water impoundments. Amer. Soc. Civil Engr., New York, NY.

To, Y. S. and C. N. Randall. 1975. The effect of sediment on reservoir water quality. Pages 590–597, in proceedings of the Second National Conference on Complete Wateruse, Amer. Inst. Chem. Eng.

U.S. Environmental Protection Agency. 1978. National Eutrophication Compedians Working Papers 474–477. Corvallis Environmental Research Laboratory and Las Vegas Environmental Monitoring and Support Laboratory, U.S. Environmental Protection Agency.

Vollenweider, R. A. 1968. The scientific basis of lake and stream eutrophication, with particular reference to phosphorus and nitrogen as eutrophication factors. Technical Report DAS/DSI/68. Organization for Economic Cooperation and Development, Paris.

Vollenweider, R. A. 1976. Advances in defining critical loading levels for phosphorus in lake eutrophication. *mem. Ist. Ital. Idrobiol.* 33:53–83.

Walker, W. W. 1982a. A simplified method for predicting phosphorus gradient potential in reservoirs, prepared for Environmental Laboratory. EWQOS Work Unit 1-E, Working paper No. 10. USAE Waterways Experiment Station, Vicksburg, MS.

Walker, W. W., W. W. 1982b. Empirical methods for predicting eutrophication in impoundments. Report 2: Model Testing, prepared for Office Chief of Engineers, U.S. Army, Washington, DC. Technical Report E-81-9. U.S. Army Corps of Engineers. Waterways Experiment Station, Vicksburg, MS.

Walker, W. W., W. W. 1985. Empirical methods for predicting eutrophication in impoundments. Report 3: Model Refinements, prepared for Office, Chief of Engineers. U.S. Army, Washington, DC. Technical Report E-81-9. U.S. Army Corps of Engineers. Waterways Experiment Station, Vicksburg, Ms.

Wright, J. C. 1967. Effect of impoundments on productivity, water chemistry, and heat budgets of rivers. Pages 188–199 in Reservoir Fishery resources. Am. Fish. Soc., Washington, DC.

Wunderlich, W. O. 1971. The dynamics of density-stratified reservoirs. Pages 219–231 in G. E. Hall, ed. Reservoir fisheries and limnology. Spec. Pub. 8th Amer. Fish Soc. Washington, DC.

Zimbalevskaya, L. N., L. A. Zhuravleva, L. A. Khoroshikh, V. I. Pugach, L. E. Kostikova, M. N. Dekhtyar, and A. B. Yakubovsky. 1976. Eutrophication of the Kremenchug Reservoir shallows. *Limnologica.* 10(2):321–324.

CHAPTER 6

Reservoir Primary Production

BRUCE L. KIMMEL, OWEN T. LIND, AND LARRY J. PAULSON

Reservoirs occupy an intermediate position between rivers and natural lakes with regard to 1) morphologic and hydrologic characteristics, 2) the relative importance of external nutrient inputs and internal nutrient cycling, and 3) the significance of allochthonous vs. autochthonous sources of organic matter to the food web (Table 6.1). Because reservoirs combine numerous features of river and lake environments, they are often described as "river-lake hybrids" (e.g., Lind 1971, Henderson et al. 1973, Ackermann et al. 1973, Ryder et al. 1974, Margalef 1975, Ryder 1978, Goldman and Kimmel 1978, Thornton et al. 1981, Jenkins 1982, Benson 1982, Groeger and Kimmel 1984, Kimmel and Groeger 1984, Carline 1986, Soballe and Kimmel 1987, Soballe et al. in press). The "river-lake hybrid" analogy encourages the use of extant knowledge of rivers, natural lakes, and river-lake interactions to form a conceptual basis for understanding 1) the spatial and temporal heterogeneity in physical, chemical, and biological conditions within individual reservoirs, 2) variations in limnological and ecological properties among reservoirs and among reservoir types, 3) differences between and similarities of man-made and natural lakes, and 4) the structure, function, and dynamics of reservoir ecosystems.

The purpose of this chapter is to discuss primary productivity in reservoir ecosystems. Our discussion focuses on the phytoplankton because of its dominance of the primary production in most large impoundments. First, we describe breifly the reservoir primary producers, the environmental factors that control phytoplankton production in reservoirs, and the relative magnitude and variability of reservoir productivity. Second,

Table 6.1 Aquatic ecosystem characteristics for which reservoirs occupy an intermediate position relative to rivers and natural lakes

Features	Rivers	Reservoirs	Lakes
Basin morphology	Elongate, channelized	Intermediate	Circular/ovoid, bowl-shaped
Flow	Rapid, directional	Intermediate	Slow, non-directional
Flushing rate	Rapid	Intermediate	Slow
Watershed influence	Very great	Intermediate	Less
Suspended particle load	High	Intermediate	Low
Nutrient supply	Advection, relatively continuous	Both	In situ recycling
Primary mode of nutrient loss	Advection	Both	Sedimentation
Organic matter supply	More allochthonous	Intermediate	More autochthonous
Spatial structure	Longitudinal gradients	Both	Vertical gradients

we discuss horizontal and vertical gradients of environmental factors that influence the distribution and magnitude of phytoplankton production within reservoirs. Third, we consider the relationship of management-oriented empirical models and our current understanding of reservoir ecosystems. Finally, we discuss factors that influence the productivity of multiple-impoundment series and the need for a river-basin-level approach to the ecology of reservoir series. We discuss these topics within the context of the "river-lake hybrid" analogy, as we believe that it provides a valid initial basis for improving our understanding of reservoirs as ecosystems.

RESERVOIR PRIMARY PRODUCERS AND ENVIRONMENTAL FACTORS CONTROLLING PRIMARY PRODUCTION

Primary Producers in Reservoirs

As in rivers and lakes, the primary (photoautotrophic) producers in reservoirs belong to one of four categories: planktonic algae (phytoplankton), planktonic phototrophic bacteria, attached algae (periphyton), and rooted macrophytes. The relative contributions of these primary producer groups to the total photosynthetic production of organic matter in reservoirs are undetermined. However, abiogenic turbidity (due to suspended silts and clays eroded from the watershed) and large water-level fluctuations (due to flood control and hydropower operations) often restrict the development of attached algal and rooted macrophyte communities in reservoirs and thus maximize the contribution of the planktonic producers to the total primary production (e.g., Ellis 1936, Isom 1971, Ryder 1978, Kimmel and Groeger 1984). Although photosynthetic bacteria are probably common in clear, stably stratified reservoirs and contribute to the development of deep chlorophyll peaks (Knowlton and Jones 1989), their contribution to reservoir primary production is almost certainly small in comparison to that of the phytoplankton.

Despite the general predominance of reservoir primary production by phytoplankton, extensive littoral macrophyte and epiphyte communities can occur in reservoirs with relatively clear water and stable water levels (Little 1966, Barko 1981). For example, luxuriant macrophyte-epiphyte communities develop in embayments of Tennessee River mainstem impoundments, in which water-level fluctuations are minimized for navigational purposes, and probably contribute significantly to the biological productivity of these and other rapidly flushed systems (Placke and Poppe 1980, Placke 1983).

Submersed timber left standing in reservoir basins to serve as game fish habitat also provides physical substrate for periphyton colonization and growth and thereby increases the surface area available for periphyton production and the development of aufwuchs communities. "Aufwuchs" refers to the entire assemblage of organisms (algae, bacteria, fungi, protists, and micro- and macroconsumers) that develops in association with submersed surfaces. Trees left standing in the Richard B. Russell Reservoir (Georgia–South Carolina) supported luxuriant periphyton growth in the photic layer of the water column as the new reservoir filled (Hains 1987). In two Missouri River impoundments, Lake Francis Case and Lewis and Clark Lake, the summer densities of macroinvertebrates were four times greater and 11 times greater, respectively, on submersed timber than on adjacent bottom sediments (Cowell and Hudson 1967, Claflin 1968). In both reservoirs benthos densities were significantly correlated with periphyton biomass (Ploskey 1981). Periphyton growth could be more significant to game fish production than is reflected by its relative contribution to reservoir primary production because it supports a localized concentration of organisms (i.e., an aufwuchs community and associated consumers) that may be exploited more efficiently by higher consumers than is the more highly dispersed plankton community.

Riverine phytoplankton and river-borne drifting periphyton can comprise a significant input of algal biomass to reservoirs (e.g., Megard 1981). However, sedimentation of large, thick-walled, rapidly-sinking riverine algae in the less turbulent reservoir environment appears to minimize the riverine contribution to the reservoir algal standing crop. Soballe and Bachmann (1984) reported high rates of algal sedimentation in the uplake region of Red Rock Reservoir, Iowa (mean annual water residence time = 11 days), negligible phytoplankton losses due to sinking near the dam, and an uplake to downlake reduction in mean algal cell size; their results thus demonstrated a rapid replacement of riverine algae by lacustrine phytoplankton.

However, the discharge of planktonic organisms from upstream impoundments, especially from those with near-surface discharges, is known to contribute to the plankton standing crop and productivity of rivers and reservoirs downstream (Brook and Rzoska 1954, Talling and Rzoska 1967, Hammerton 1972, Shiel and Walker 1984, Petts 1984). The viability of phytoplankton in reservoir discharges depends on the outlet depth. Surface or near-surface discharges export viable plankton downstream, but hypolimnial discharges release primarily moribund cells and organic detritus to downstream systems (Coutant 1963, Cowell 1970, Lind 1971, Stroud and Martin 1973).

Surprisingly little data exist on the relative importance of the advective input of phytoplankton as compared to autochthonous production (Soballe and Bachmann 1984). To explain the observed level of fish production in a rapidly flushed mainstem impoundment of the Tennessee River (Nickajack Reservoir, Tennessee, 3-d water residence time), Adams et al. (1983) hypothesized that viable phytoplankton exported from a larger, more productive reservoir (Chickamauga Reservoir, Tennessee, 7-d water residence time) upstream provided an important supplement to the autochthonous algal production in the reservoir downstream. However, this hypothesis remains untested. The importance of biotic linkages between upstream and downstream reservoirs is an area in need of quantitative study.

Size Distribution of Phytoplankton Productivity in Reservoirs

By virtue of their semifluvial nature and the relatively high levels of turbulence and nutrient loading in reservoirs, large phytoplankton (>20 μm) might be expected to be more important in reservoirs than in natural lakes (Kalff and Knoechel 1978, Malone 1980, Watson and Kalff 1981). However, comparisons of the size distribution of planktonic autotrophy (i.e., as indicated by photosynthetic ^{14}C uptake) in a variety of reservoirs of differing trophic status, turbidity, and productivity levels indicate that small algae (<8 μm) dominate phytoplankton productivity in reservoirs (Kimmel 1983, Kimmel and Groeger 1987), as they do in most lacustrine environments and in the open ocean (Stockner and Antia 1986, Platt and Li 1986).

For example, in DeGray Reservoir (Arkansas), usually 50–75% of the total photosynthetic ^{14}C uptake was attributable to phytoplankton that passed through an 8-μm Nucleopore polycarbonate filter. Similarly, free-living (<1.0-μm) bacterioplankton, rather than bacterial aggregates or bacteria attached to suspended particles, accounted for 75–90% of the planktonic microheterotrophy (Kimmel and Groeger 1987) in DeGray Reservoir. Together these data from reservoirs show that over a broad range of environmental conditions the predominant fractions of planktonic autotrophy and microheterotrophy are associated with <8-μm algae and <1-μm bacteria, respectively. Furthermore, these results support the view that pelagic ecosystem metabolism is dominated by very small organisms (Pomeroy 1974, Sieburth et al. 1978, Williams 1981, Ducklow 1983, Stockner and Antia 1986, Platt and Li 1986) and demonstrate that such a view applies not only to oligotrophic open-ocean, coastal, and lacustrine environments but also extends to more productive lakes and reservoirs.

Factors Controlling Phytoplankton Production in Reservoirs

Reservoir phytoplankton productivity and biomass levels are dependent on several interrelated physical, chemical, and biological factors, that are themselves functions of the climatic and hydrologic regimes, the size and nature of the watershed, reservoir basin morphology, nature and volume of river inflow, and the reservoir food-web structure. Longitudinal changes in reservoir basin morphology and flow velocity result in longitudinal differences in flushing rates, mixing depths, suspended particle concentrations, nutrient levels, and thus in light and nutrient availability in various parts of the reservoir. These interactions will be discussed in detail below with regard to gradients in reservoirs.

Phytoplankton productivity and biomass in reservoirs are controlled by the same energy and nutrient inputs (Brylinsky and Mann 1973, Schindler 1978, Brylinsky 1980) and balance of gains and losses (Jassby and Goldman 1974, Kalff and Knoechel 1978, Westlake et al. 1980) that govern other planktonic systems. The basic factors controlling phytoplankton productivity (temperature, light, macro- and micronutrient availability) have been reviewed extensively by others (e.g., Talling 1961, 1971, Lund 1965, Goldman 1968, Fogg 1975, Steeman Nielsen 1975, Harris 1978, Westlake et al. 1980, Reynolds 1984, Harris 1986) and will not be elaborated again here. In brief, however, algal growth rate is determined by the relative availability of some limiting factor (e.g., light) and increases with an increase in that rate-limiting factor until another factor (e.g., nutrient availability or temperature) becomes more limiting. However, because the planktonic environment is physically, chemically, and biotically dynamic, the concept of a complex of environmental factors controlling algal growth is more appropriate than that of control by a single limiting factor (e.g., O'Brien 1972, 1974, Harris 1978, 1980a, 1988).

The accumulation of algal biomass, the end product of phytoplankton growth, is determined both by the rate of biomass production and the rate of biomass loss. Two categories of controlling factors affect these rates: 1) physical and chemical factors that influence the availability of light and nutrients for photosynthesis and algal growth, temperature-dependent metabolic rates, and phytoplankton losses due to cell sedimentation and washout; and 2) biotic factors influencing algal photosynthetic efficiency, extracellular excretion of photosynthate, and phytoplankton losses resulting from predation or parasitism. Thus, fluctuations in the phytoplankton standing crop reflect changes in the net balance of several gain and loss processes, often summarized as:

$$B_t = (A_i + \mathrm{GP}) - (A_o + R + G + S + M)$$

where B_t = phytoplankton biomass change over some time interval t
A_i = advective input
GP = gross phytoplankton production
A_o = advective output
R = respiration

G = grazing
S = sinking
M = other sources of mortality or loss.

See Jassby and Goldman (1974) and Crumpton and Wetzel (1982) for detailed considerations of phytoplankton community dynamics by the evaluation of algal gain and loss rates.

Nutrient control of phytoplankton production is often emphasized in investigations of marine and freshwater planktonic systems (e.g., Goldman 1968, Likens 1972, Vollenweider 1976, Schindler 1978). Rivers have been less throughly studied than lakes, but it is generally recognized that limited light availability (due to high suspended solids concentrations and rapid vertical mixing) is often a more important controlling factor in rivers (e.g., Mann et al. 1972, Naiman and Sedell 1981, Wetzel 1975a). Two primary factors controlling phytoplankton productivity in reservoirs, light and nutrient availability, are themselves functions of inflow characteristics (in particular, the suspended sediment and dissolved nutrient loads) and the vertical mixing regime.

The residence time of water in lakes, rivers, and reservoirs exerts an important influence on phytoplankton biomass and production levels (Soballe and Kimmel 1987). As in a chemostat, if the flushing rate exceeds the phytoplankton doubling rate, the accumulation of biomass and therefore algal productivity can be limited by advective losses (washout) of cells (e.g., Brook and Woodward 1956, Dickman 1969, Javornicky and Komarkova 1973, Straskraba and Javornicky 1973). Growth rates of algae in laboratory culture range from about 0.25 to 2.0 days per doubling (Hoogenhout and Amesz 1965); however, phytoplankton growth rates in nature are usually much slower. Westlake (1980) reported a range of phytoplankton doubling rates from 0.12 to 7.5 days at the depth of optimal photosynthesis. The water residence time at which algal biomass accumulation can be controlled by washout is about one week (Uhlmann 1968). Numerous run-of-the-river mainstem and re-regulation impoundments have average retention times of less than seven days, and therefore, phytoplankton production and species composition may often be influenced by flushing rate in such systems. Both phytoplankton and zooplankton biomass were negatively correlated with flushing rate in re-regulation reservoirs (mean retention time ca. 1.5 days) of the Vltava Cascade reservoir system in Czechoslovakia (Straskraba and Javornicky 1973).

Phytoplankton washout can also be important in lakes and reservoirs with intermediate water residence times when the flushing rate of the photic zone is high (e.g., during spring periods of high inflow) (Carmack et al. 1979). However, if the flushing rate does not exceed the mean doubling time of the phytoplankton assemblage, increased inflow may enhance phytoplankton productivity by increasing nutrient availability. Turner et al. (1983) reported that increased inflow reduced the water residence time from 44 to 22 days but resulted in a doubling of the annual phytoplankton production in Lake Tahlquin (Florida).

It has been hypothesized that there are threshold values for advective or water residence time effects in both lotic and lentic ecosystems (Margalef 1960, Ford and Thornton 1979, Meyer and Likens 1979) and that significant changes in system structure and function are likely if advection exceeds these thresholds. A recent statistical comparison of factors influencing phytoplankton abundance in rivers, lakes, and impoundments (Soballe and Kimmel 1987) supports this hypothesis and demonstrates that threshold effects of water residence time on the algal standing crop are detectable at both seasonal and whole-system scales of resolution. Simulation models (Pridmore and McBride 1984, Soballe and Threlkeld 1985), empirical studies (Brook and Woodward 1956, Talling and Rzoska 1967, Dickman 1969, OECD 1982), and statistical comparisons (Soballe and Kimmel 1987) suggest that the direct advective effects of water renewal on phytoplankton abundance are restricted to systems with water residence times of less than 60–100 days.

Traditionally, limnologists attribute the control of phytoplankton productivity and community composition primarily to physical and chemical factors (e.g., Wetzel 1983, Harris 1980, 1986); however, biotic factors (e.g., food-web interactions) can also affect primary production levels (Hrbacek et al. 1961, Shapiro 1980, Carpenter et al. 1985, 1987). The cascading trophic-interactions hypothesis (Carpenter and Kitchell 1984, Carpenter et al. 1985) suggests that lake ecosystem productivity is regulated hierarchically through both biotic and abiotic mechanisms and that much of the unexplained variance in lake productivity (see Schindler 1978) results from differences in food-web structure and trophic interactions. Thus, differences in piscivory "cascade" down the food web, are expressed in the population dynamics and behaviors of zooplanktivorous fishes, in their effects on herbivorous prey, and ultimately, through both selective grazing and nutrient cycling, in the composition and productivity of the phytoplankton community.

Carpenter et al. (1987) performed whole-lake manipulations of fish populations to test the hypothesis that higher trophic levels regulate zooplankton and phytoplankton community structure, biomass, and primary productivity. Experimental piscivore addition and planktivore removal resulted in major changes in both plankton community composition and phytoplankton production in the manipulated lakes, and it was concluded that abiotic factors and food-web effects were equally important regulators of lake productivity. Although considerable polarization has developed between advocates of "bottom-up" control (i.e., by nutrients, light, mixing) vs. "top-down" control (by food-web interactions) of lentic productivity, it seems reasonable to assume that both abiotic and biotic controls are important at various times in most lakes and reservoirs and that their relative importance will vary temporally, spatially, and between systems (e.g., O'Neill et al. 1986, Bartell et al. 1988).

The Magnitude and Variability of Reservoir Phytoplankton Productivity

Estimates of the mean daily phytoplankton production for a number of temperate and tropical reservoirs are summarized in Table 6.2. The reservoirs listed in Table 6.2 span the trophic range from oligotrophy to eutrophy, and their productivity levels fall within the same range as those of natural lakes (Table 6.3). However, as a group, reservoirs appear to be somewhat more productive than natural lakes. Of 102 natural lakes listed by Wetzel (1983) and Brylinsky (1980), 46% are oligotrophic, 40% are mesotrophic, and only 14% are eutrophic. In contrast, only 16% of the 64 reservoirs listed in Table 6.2 are oligotrophic, 52% are mesotrophic, and 33% are eutrophic. These differences in relative productivity probably result from differences in watershed size, watershed fertility, and water-residence times between lakes and reservoirs. Most reservoirs have significantly higher drainage-area-to-lake-surface-area ratios, greater external nutrient loading, and shorter water residence times than do most natural lakes (Thornton et al. 1981, Benson 1982, Kimmel and Groeger 1984, Soballe and Kimmel 1987).

At present we do not know of an ultra-oligotrophic (<50 ug C m-2 day-1) reservoir. It is interesting to note that in the most "oligotrophic" reservoir of Table 6.2 (Tuttle Creek Reservoir, Kansas) phytoplankton productivity is severely light-limited by high levels of abiogenic turbidity (Marzolf and Osborne 1971, Marzolf Chapter 7) rather than nutrient-limited, as is normally implied by the trophic-status nomenclature.

Multiple-year data that permit evaluation of year-to-year variability in

Table 6.2 Summary of reservoir phytoplankton productivity estimates. Productivity values, listed in increasing order, represent the mean daily production for the entire year or growing season, unless noted otherwise. Asterisks indicate average values for two or more annual production estimates. Trophic state categories, as indicated by ^{14}C-method estimates, are those of Likens (1975) and Wetzel (1983)

Reservoir, Location	Year	Mean daily production ($mg\ C\ m^{-2}\ d^{-1}$)	Comments	Reference
OLIGOTROPHIC: 50–300 $mg\ C\ m^{-2}d^{-1}$				
Tuttle Creek, Kansas	1970, 71	67	^{14}C; highly turbid, light-limited system	Marzolf and Osborne (1971)
Koocanusa, Montana	1972–75	84*	^{14}C	Woods (1981)
Sam Rayburn, Texas	1977–78	102	^{14}C	Lind (1979)
Merle Collins, California	1965–68	106*	^{14}C; new reservoir, productivity increased ca. 33% yr^{-1} from 1965 to 1968	Chamberlain (1972)
Smallwood, Labrador, Canada	1974, 75	138*	^{14}C	Ostrofsky (1978), Ostrofsky and Duthie (1978)
Canyon, Texas	1976	184	^{14}C	Hannan et al. (1981)
Folsom, California	1965, 66	189*	^{14}C	Chamberlain (1972)
DeGray, Arkansas	1979, 80	199*	^{14}C	Kennedy (unpubl. data)
Nottely, Georgia	1969	208	^{14}C	Taylor (1971)
Nickajack, Tennessee	1973	235	^{14}C, summer estimates	Placke and Poppe (1980)
MESOTROPHIC: 250–1,000 $mg\ C\ m^{-2}d^{-1}$				
Francis Case, South Dakota	1968	260	Net O_2 change, summer estimates	Martin and Novotny (1975)

Broken Bow, Oklahoma	1979–80	309	^{14}C, summer estimates	Kimmel (unpubl. data)
Atwood, Ohio	1949	339	Gross O_2 change, April–June	Wright (1954)
Norris, Tennessee	1967	360	^{14}C	Taylor (1971)
Kuibyshev, USSR	—	360–780	?, ca. May–October	Salmanov and Sorokin (1972), in Poddubny (1976)
Monroe, Indiana	1975, 76	378*	^{14}C, April–June	Chang and Frey (1977), Santiago (1978)
Rybinsk, USSR	1955, 58–61, 64–72	380*	^{14}C, May–November	Romanenko (1978)
Mahinerangi, New Zealand	1964–66	76	^{14}C, undergoing cultural eutrophication	Mitchell and Galland (1981)
	1968–70	210		
	1976–79	380		
Gorky, USSR	1956	406	^{14}C, May–October	Sorokin et al. (1959), in Poddubny (1976)
Isabella, Michigan	1977–78	424	Net O_2 change, seasonal estimates	Groeger (1979)
Normandy, Tennessee	1982, 83	458*	^{14}C	Kimmel (unpubl. data)
Summersville, West Virginia	1971, 72	466	^{14}C	Fraser (1974)
Slapy, Czechoslovakia	1962–67	501*	Gross O_2 change, April–September	Javornicky and Komarkova (1973)
North Lake, Texas	1976	521	^{14}C	Stuart and Stanford (1979)
Klicava, Czechoslovakia	?	523	?	Brylinsky (1980)
Lewis and Clark, Nebraska	1968	530	Net O_2 change, summer estimates	Martin and Novotny (1975)

Table 6.2 (*continued*)

Reservoir, Location	Year	Mean daily production (mg C m^{-2} d^{-1})	Comments	Reference
Bratsk, USSR	?	577	?	Brylinsky (1980)
Keowee, South Carolina	1973, 74	582*	^{14}C	Dillon and Rogers (1980)
Par Pond, South Carolina	1969–73	606*	^{14}C, reactor cooling reservoir	Tilly (1975)
Hebgen, Montana	1965	658	Net O_2 change, summer estimates	Martin and Arneson (1978)
Hartwell, South Carolina	1971	660	^{14}C	Abernathy and Bungay (1972)
West Point, Alabama-Georgia	1977	689	^{14}C	Davies et al. (1980)
Cheat, West Virginia	1971	695	^{14}C	Volkmar (1972)
Kiev, USSR	1967	708	Net O_2 change, April–October	Gak et al. (1972)
Lipno, Czechoslovakia	?	717	?	Brylinsky (1980)
Doe Valley, Kentucky	1969–72	729*	^{14}C	Bacon (1973)
Ivan'kovskii, USSR	1956	733	?, May–September	Pyrina (1966), in Poddubny (1976)
Lake Mead, Arizona–Nevada	1977–78	810	^{14}C	Paulson et al. (1980)
Waco, Texas	1968, 1977–78	814*	^{14}C	Kimmel and Lind (1972), Lind (1979)
Bighorn, Montana–Wyoming	1968–70	827*	Estimated from chlorophyll and light data after Ryther and Yentsch (1957)	Soltero and Wright (1975)

Lake Talquin, Florida	1972–74	833*	^{14}C	Turner et al. (1983)
Pena Blanca, Arizona	1959–61	899	Gross pelagic production by O_2 change and estimates from chlorophyll data	McConnell (1963)
Texoma, Oklahoma–Texas	1979, 80	934	^{14}C, summer estimates	B. L. Kimmel (unpubl. data)
Douglas, Tennessee	1969	940	^{14}C	Taylor (1971)
EUTROPHIC: >1000 mg C $m^{-2}d^{-1}$				
Canyon Ferry, Montana	1958	1125	Net O_2 change, April – September	Wright (1958, 1959, 1960)
Chickamauga, Tennessee	1977	1286	^{14}C, summer estimates	Placke and Poppe (1980)
Moss, Texas	1976	1302	^{14}C	Silvey and Stanford (1978)
Cherokee, Tennessee	1969	1416	^{14}C	Taylor (1971)
Mohave, Arizona–Nevada	1976–78	1420*	^{14}C	Priscu et al. (1982), Paulson et al. (1980)
Kentucky, Kentucky–Tennessee	1968	1440	^{14}C	Taylor (1971)
Krishnagiri, India	——	1601	Net O_2 change	Sreenivasan (1972)
Beech, Tennessee	1968	1619	^{14}C	Taylor (1971)
Wagon Train, Nebraska	1969, 70	1781*	^{14}C, summer estimates, 726 and 2836 mg C $m^{-2}d^{-1}$ in 1969 and 1970, respectively. Differences due to high inorganic turbidity in 1969 relative to 1970.	Anderson and Hergenrader (1973)

Table 6.2 (*continued*)

Reservoir, Location	Mean daily production			Reference
	Year	($mg\ C\ m^{-2}\ d^{-1}$)	Comments	
Ashtabula, North Dakota	1966–68	1828*	Net O_2 change	Peterka and Reid (1966), Knutson (1970), cited in Soltero et al. (1975)
Long Lake, Washington	1972, 73	1903	Estimated from chlorophyll and light data after Ryther and Yentsch (1957), July–March	Soltero et al. (1975)
Tjeukemeer, Netherlands	1969	2055	^{14}C, Dutch polder reservoir	Beattie et al. (1972)
Big Eau Pleine, Wisconsin	1975, 76	2145*	^{14}C, April–November	Sullivan (1978)
Sathanur, India	——	2312	Net O_2 change	Sreenivasan (1972)
Sandynulla, India	——	2319	Net O_2 change	Sreenivasan (1972)
Stanley, India	——	2329	Net O_2 change	Sreenivasan (1972)
Bhavanisagar, India	——	2329	Net O_2 change	Sreenivasan (1972)
Kainji, Nigeria	1970, 71	2434	Gross O_2 change	Lelek (1973)
Volta, Ghana	1966	2547	Gross O_2 change, 2.0-5.2 $g\ C\ m^{-2} d^{-1}$ range within reservoir	Viner (1970)
Amaravathi, India	——	3230	Net O_2 change	Sreenivasan (1972)
Stagecoach, Nebraska	1969, 70	3975*	^{14}C, summer estimates	Anderson and Hergenrader (1973)

Table 6.3 Comparison of primary productivity ranges for: (A) regional aquatic ecosystem types (after Likens 1975), (B) lakes listed by Wetzel (1983) and Brylinsky (1980), and (C) reservoirs listed in Table 6.2

Aquatic ecosystems	$mg\ C\ m^{-2}\ d^{-1}$ [a]	Reference
(A) Tropical lakes	100–7600	Likens (1975)
Temperate lakes	5–3600	
Arctic lakes	1–170	
Antarctic lakes	1–35	
Alpine lakes	1–450	
Temperate rivers	1–3000	
Tropical rivers	1–?	
(B) 102 natural lakes	3–5529[b]	Wetzel (1983), Brylinsky (1980)
(C) 64 reservoirs in Table 2	67–3975	This paper

[a]Averaged over the estimated growing season.
[b]Conversion factors: 40 kJ/g carbon, 245 growing season d/year.

lake and reservoir productivity are rare. However, the comparisons that are possible (Table 6.4) suggest that phytoplankton production in reservoirs is not significantly more variable on an annual basis than is phytoplankton production in natural lakes. However, if finer-scale (e.g., daily) productivity data were available, productivity in most reservoirs would likely be more variable than in most natural lakes as a result of 1) more rapid flushing rates and 2) the continued influence of inflow from the impounded river.

Some ecosystems are "stable" by virtue of an absence of perturbing forces; however, they lack resilience and perturbations may result in significant ecosystem level changes. Reservoirs, which are known for their dynamic and heterogeneous nature, experience frequent perturbations due to both natural meteorologic and hydrologic variations and reservoir operations. The relatively low variability of annual phytoplankton production in reservoirs suggests that reservoir ecosystems possess a resilience or "dynamic stability" (Barnes and Mann 1980, Webster et al. 1985) that tends to return them to their previous state in spite of repeated perturbations. The inherent resiliency exhibited by many reservoir ecosystems is probably largely a result of 1) relatively high external nutrient loading and 2) relatively short water residence times.

Table 6.4 Variability of annual phytoplankton productivity in natural lakes and reservoirs for which at least four years of production data were available. Phytoplankton productivity is expressed either as total annual production (g C m^{-2} yr^{-1}) or as the mean daily production for a year or growing season (mg C m^{-2} d^{-1}). Relative variation is expressed as the coefficient of variation (CV = SD/$\bar{X}$).

Aquatic system and data source	Years of measurement (n)	Phytoplankton productivity ($\bar{X}$ ± SD) g C m^{-2} yr^{-1}	mg C m^{-2} d^{-1}	(CV%)
Natural lakes				
Lawrence Lake, Mich.[a]	1968–74 (7)	41.1 ± 5.4	112.6 ± 14.9	13.2
Castle Lake, Calif.[b]	1960, 63, 67–71, 73 (8)	43.6 ± 15.2	—	34.9
Lake Myvatn, South Basin, Iceland[c]	1971–76 (6)	117.8 ± 13.8	—	11.7
Lake Esrom, Denmark[d]	1955–57, 61–72 (14)	237.6 ± 50.9	—	21.2
Lake Vechten, Nether lands[e]	1969, 70, 72–74, 77–80 (9)	—	280.0 ± 98.9	35.3
Lake Kinneret, Israel[f]	1969–72 (4)	—	1423.5 ± 365.7	25.7
			[Natural lake $\overline{CV}$ ± SD = 23.7 ± 10.2]	
Reservoirs				
Koocanusa, Mont.[g]	1972–75 (4)	—	84 ± 20	23
Merle Collins, Calif.[h]	1965–68 (4)	—	106 ± 40	38
Rybinsk, USSR[i]	1955, 58–61, 64–72 (13)	—	380 ± 182	48
Slapy, Czechoslovakia[j]	1962–67 (6)	—	501 ± 105	21
Par Pond, So. Car.[k]	1967–73 (5)	—	647 ± 133	21
			[Reservoir $\overline{CV}$ ± SD = 30.2 ± 12.2]	

[a]Wetzel (1983), [b]C. R. Goldman (unpublished data). [c]Jonasson and Adalsteinsson (1979). [d]Jonasson (1977), [e]Kloet (1982), [f]Berman and Pollingher (1974), [g]Woods (1981). [h]Chamberlain (1972), [i]Romanenko (1978), [j]Javornicky and Komarkova (1973), [k]Tilly (1975).

GRADIENTS WITHIN RESERVOIRS — EFFECTS ON PHYTOPLANKTON PRODUCTION

Lotic ecosystems (streams, rivers) are characterized by longitudinal gradients in channel morphology, flow velocity, water temperature, substrate type, and biotic communities (e.g., Hynes 1970, 1975, Cummins 1974, 1979, Vannote et al. 1980, Minshall et al. 1983). Vertical gradients of light, temperature, dissolved substances, and production and decomposition processes are primary features of lentic (lake-like) environments (e.g., Hutchinson 1957, Wetzel 1983). By virtue of their river-lake hybrid nature, reservoirs possess both horizontal and vertical gradients in the abiotic factors that control phytoplankton production.

Longitudinal Gradients

Reservoirs exhibit a marked degree of spatial heterogeneity in phytoplankton productivity and biomass due to longitudinal gradients in basin morphology, flow velocity, water residence time, suspended solids, and light and nutrient availability. Typically, three zones are distinguishable along the longitudinal axis of a typical reservoir (Figures 6.1 and 6.2):

1) The uplake *riverine zone* is a lotic environment. It is characterized by higher flow, shorter water residence time, and higher levels of available nutrients, suspended solids, and light extinction relative to the downstream portions of the reservoir. Abiogenic turbidity often limits light penetration and thereby limits the thickness of the photic layer. Although phytoplankton biomass and productivity per unit volume (mg C m^{-3}) of the photic zone can be high in the riverine zone, the depth of the mixed layer (Z_m) usually exceeds that of the photic layer (Z_p) and consequently, the areal primary productivity (mg C m^{-2}) is often light-limited.
2) The *transition zone* is characterized by higher phytoplankton productivity and biomass occurring in conjunction with increasing basin breadth, decreasing flow velocity, increased water residence time, sedimentation of silt and clay particles from near-surface waters, and increased light penetration. The transition zone is often associated with the plunge point (see Chapters 2 and 5), if one occurs. Because both light and nutrients are available for algal photosynth-

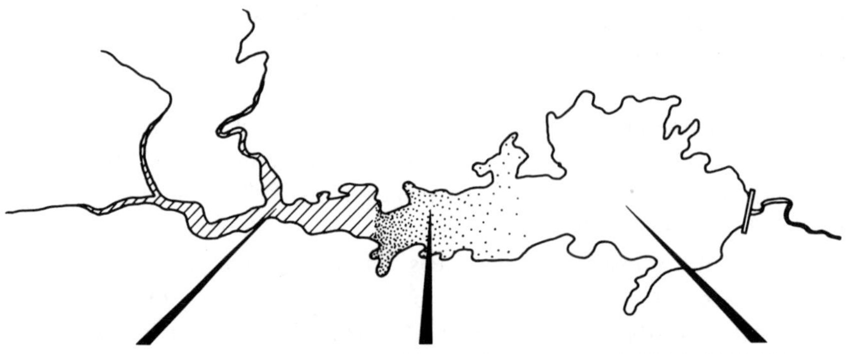

RIVERINE ZONE	TRANSITIONAL ZONE	LACUSTRINE ZONE
• NARROW, CHANNELIZED BASIN	• BROADER, DEEPER BASIN	• BROAD, DEEP, LAKE-LIKE BASIN
• RELATIVELY HIGH FLOW	• REDUCED FLOW	• LITTLE FLOW
• HIGH SUSP. SOLIDS, TURBID, LOW LIGHT AVAIL., Zp < Zm	• REDUCED SUSP. SOLIDS, LESS TURBID, LIGHT AVAIL. INCREASED	• REL. CLEAR, LIGHT MORE AVAIL. AT DEPTH, Zp > Zm
• NUTRIENT SUPPLY BY ADVECTION, REL. HIGH NUTRIENTS	• ADVECTIVE NUTRIENT SUPPLY REDUCED	• NUTRIENT SUPPLY BY INTERNAL RECYCLING, REL. LOW NUTRIENTS
• LIGHT-LIMITED PPR	• PPR/m^3 REL. HIGH	• NUTRIENT-LIMITED PPR
• CELL LOSSES PRIMARILY BY SEDIMENTATION	• CELL LOSSES BY SEDIMENTATION AND GRAZING	• CELL LOSSES PRIMARILY BY GRAZING
• ORGANIC MATTER SUPPLY PRI-ALLOCHTHONOUS, P < R	• INTERMEDIATE	• ORGANIC MATTER SUPPLY PRIMARILY AUTOCHTHONOUS, P > R
• MORE "EUTROPHIC"	• INTERMEDIATE	• MORE "OLIGOTROPHIC"

Figure 6.1 Longitudinal zonation in environmental factors controlling light and nutrient availability for phytoplankton production, algal productivity and standing crop, organic matter supply, and trophic status in an idealized reservoir. Modified from Kimmel and Groeger (1984).

esis, the transition zone can be the most fertile region in the reservoir.

3) The *lacustrine zone* occurs down reservoir, nearest the dam, and usually has a longer water residence time, lower concentrations of dissolved nutrients and suspended abiogenic particles, higher water transparency, and a deeper photic layer. However, the volumetric phytoplankton productivity (mg C m^{-3}) of the photic zone is reduced (often nutrient-limited) during most of the growing season and is supported primarily by in situ nutrient cycling rather than by advected nutrients.

The relative fertility of the mixed layer generally decreases down-reservoir as the advected nutrient supply is reduced with increasing dis-

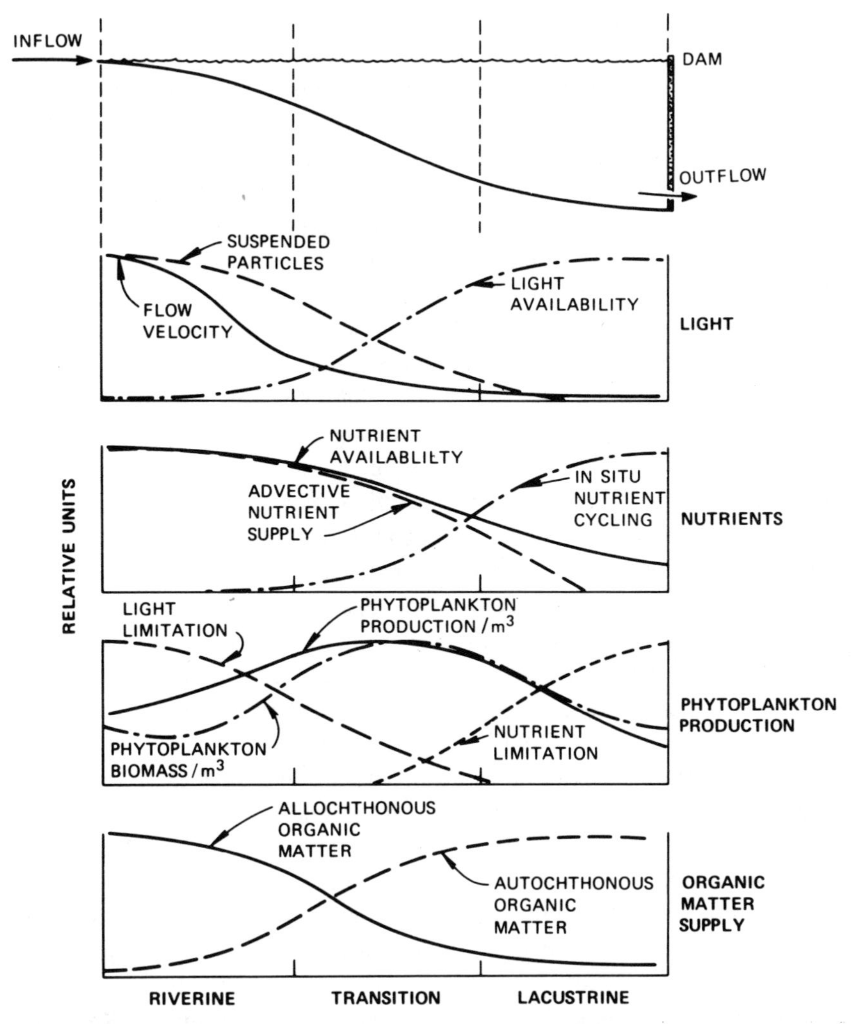

Figure 6.2 A cross-sectional view of gradients in environmental factors affecting phytoplankton productivity and biomass, and of the relative importance of allochthonous and autochthonous organic matter along the longitudinal axis of an idealized reservoir.

tance from tributary inflow points and phytoplankton production becomes more dependent on in situ nutrient regeneration (Figures 6.1 and 6.2). "Trophic state" (as reflected by Secchi depth, phosphorus supply, chlorophyll content, phytoplankton productivity, dissolved oxygen depletion, or indices based on these parameters) usually shifts from more eutrophic to more oligotrophic conditions along the riverine-transition-lacustrine gradient (e.g., Soltero and Wright 1975, McCullough 1978, Hannan et al. 1981, Placke 1983, Thornton et al. 1982, Kimmel and Groeger 1984).

The riverine, transition, and lacustrine zones within a reservoir are not discrete, invariable entities but result from the combined effects of a number of overlapping gradients (Figure 6.2). It is important to recognize that these zones are usually quite dynamic and expand and contract in response to watershed runoff events, density flow characteristics, and reservoir operating schedules. Indeed, the dynamic nature of reservoir inflow and discharge explains a great deal about why individual reservoirs can appear to be so different from each other.

Depending on watershed and inflow characteristics and flushing rates all three zones may not always be distinguishable within a particular impoundment (Figure 6.3). For example, in rapidly flushed, run-of-the-river impoundments receiving turbid inflow, conditions characteristic of the riverine zone may persist throughout most of the reservoir (e.g., Tuttle Creek Reservoir, Kansas, Marzolf and Osborne 1971, Marzolf 1981, Marzolf Chapter 7; Red Rock Reservoir, Iowa, Soballe 1981, Soballe and Bachman 1984, Kennedy et al. 1981, Thornton, Chapter 3). In contrast, a long-residence-time tributary-storage reservoir located in a relatively nonerodible nutrient-poor watershed may receive little suspended sediment and nutrient loading from tributaries, light limitation of phytoplankton producivity may be uncommon, and both the riverine and transition zones may be compressed into a small, uplake portion of the basin (e.g., DeGray Reservoir, Arkansas, Thornton et al. 1981; Norris Reservoir, Tennessee, Kimmel et al. unpublished data).

Longitudinal gradients in ecologically important physical-chemical conditions in reservoirs result in corresponding biological-physiological gradients in the reservoir phytoplankton. For example, the mixed-layer phytoplankton community in the lacustrine zone of Normandy Reservoir (a tributary-storage impoundment on the upper Duck River in south-central Tennessee) becomes nitrogen-deficient in late summer as a result of its isolation from riverine nutrient inputs and from the relatively nutrient-rich hypolimnion (Groeger and Kimmel 1988). Marked longitudinal gradients develop throughout the reservoir in water transparency, intensity of thermal stratification, and mixed-layer nutrient concentrations and are paralleled by gradients in phytoplankton productivity, algal standing crop, and physiological symptoms of algal nitrogen deficiency (e.g., lipid synthesis rates, NH_4-enhanced carbon uptake in the dark,

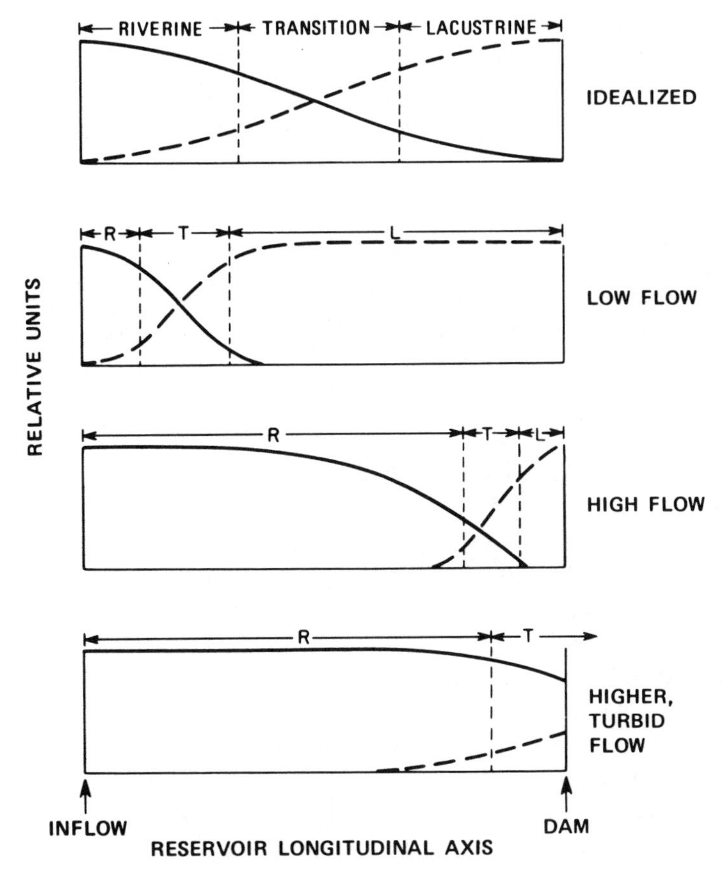

Figure 6.3 Variation in the longitudinal zonation of environmental conditions within reservoir basins. Solid and dashed lines represent the prevalence of riverine and lacustrine conditions, respectively. The longitudinal zonation is dynamic spatially and temporally and fluctuates in response to watershed runoff, inflow characteristics, density-flow behavior, and reservoir operations.

effects of nutrient enrichment on patterns of photosynthetic carbon allocation).

Although longitudinal zonation in physical, chemical, and biological factors is a distinctive characteristic of reservoir ecosystems, it is not a feature unique to reservoirs. Similar longitudinal patterns are known from elongated natural lakes (e.g., Gascon and Leggett 1977, Peters 1979, Carmack et al. 1979) and riverine estuaries (Stross and Stottlemeyer 1965). The longitudinal superposition of light and nutrient availability gradients and the consequent zonation in phytoplankton productivity may prove to be a general characteristic of semifluvial systems that receive relatively turbid nutrient-rich inflows (Kimmel and Groeger 1984).

Other Sources of Horizontal Spatial Heterogeneity

The dendritic nature of reservoir basins, tributaries of differing water quality, and embayments having limnological characteristics unique from adjacent main-channel areas result in lateral variations in productivity and thereby further contribute to the spatial heterogeneity of reservoir ecosystems. Because water masses are relatively isolated in embayments during periods of increasing pool level (see Chapter 2), embayments also provide spatial refuges from natural and man-induced perturbations associated with mainstream flows, as do backwater areas in lotic systems.

Reservoir embayments, relative to open-water regions, may be quite important to reservoir game-fish and forage-fish populations as food sourves and nursery areas, especially early in the growing season and in rapidly flushed impoundments where embayment vs. main-channel differences should be maximized (Van Winkle et al. 1981). Differences in primary and secondary productivity between reservoir embayments and main-channel areas may be further enhanced if macrophyte stands occur in the embayments. Macrophytes can serve as a point of concentration of both food and habitat for organisms at all food-web levels; thus, macrophyte beds in embayments may function as isolated pockets of highly concentrated primary and secondary production, high-quality habitat, and intensive trophic (predator-prey) interactions. Surprisingly, little research has been conducted to define the role of reservoir embayments in determining reservoir productivity.

Vertical Gradients

Photosynthetically active solar radiation (ca. 400–700 nm) decreases exponentially with depth at a rate dependent on the extent of light adsorption and scattering by dissolved and suspended matter in the water column. Regardless of nutrient availability, photoautotrophic production can occur only in the photic layer, that portion of the water column in which there is sufficient light energy to support algal photosynthesis in excess of respiration ($P > R$). The compensation depth (Z_c, where $P =$

R) is usually considered to occur at 1% incident surface light and to define the lower boundary of the photic zone (i.e., $Z_c = Z_p$). In reservoirs, as in other planktonic environments, the occurrence or nonoccurrence of stable water-column stratification and the interactions between the vertical mixing regime and light and nutrient availability have important consequences for the physiological status, productivity, size structure, and taxonomic composition of producer and consumer communities (e.g., Sheldon and Parsons 1967, Sheldon et al. 1972, 1977, Kerr 1974, Harris 1978, 1980b, Reynolds 1984).

In highly transparent reservoirs, the depth of the photic layer (Z_p) may exceed that of the mixed layer (Z_m), and phytoplankton production can occur both in the epilimnion and in euphotic portions of the metalimnion and hypolimnion (Figure 6.4*a*). Commonly, by midsummer the mixed layer becomes nutrient depleted, and metalimnetic or hypolimnetic peaks in phytoplankton biomass and productivity develop. In stratified reservoirs deep chlorophyll peaks can be formed by the accumulation of viable cells settled from the mixed layer, by active growth of cells adapted to the low light — high nutrient environment of the deeper layers, and/or by the subsurface transport of phytoplankton entrained from more productive surface waters upreservoir. Deep chlorophyll peaks have received much research attention in the oceans (e.g., Steele 1964, Lorenzen 1967, Anderson 1969, Kiefer et al. 1976) and in natural lakes (e.g., Findenegg 1964, Goldman 1968, Schindler and Holmgren 1971, Kiefer et al. 1972, Fee 1976, Brooks and Torke 1977, Richerson et al. 1978). Although they have not received comparable attention in man-made impoundments, deep chlorophyll peaks are not uncommon in transparent, shallowly stratified reservoirs (e.g., see Elser and Kimmel 1985a, Kimmel and Groeger 1987, Groeger and Kimmel 1988, Knowlton and Jones 1989).

If $Z_p = Z_m$, phytoplankton circulate through a vertical light gradient ranging from high, potentially photoinhibiting light intensities near the surface to the compensation level (where photosynthesis = respiration) at Z_p (Figure 6.4*b*). It is often assumed that mixed-layer phytoplankton adapt to the average light intensity encountered (Jewson and Wood 1975); however, photochemical mechanisms may permit phytoplankton to shift light adaptation states rapidly and thereby maximize their photosynthetic efficiency even in highly variable light fields (Vincent 1979, 1980a, Elser and Kimmel 1985a). Marra (1978) and Gallegos et al. (1980) suggested that a variable light environment enhances areal phytoplankton productivity by avoiding the effects of photoinhibition.

In turbid, well-mixed reservoirs Z_m often exceeds Z_p and planktonic

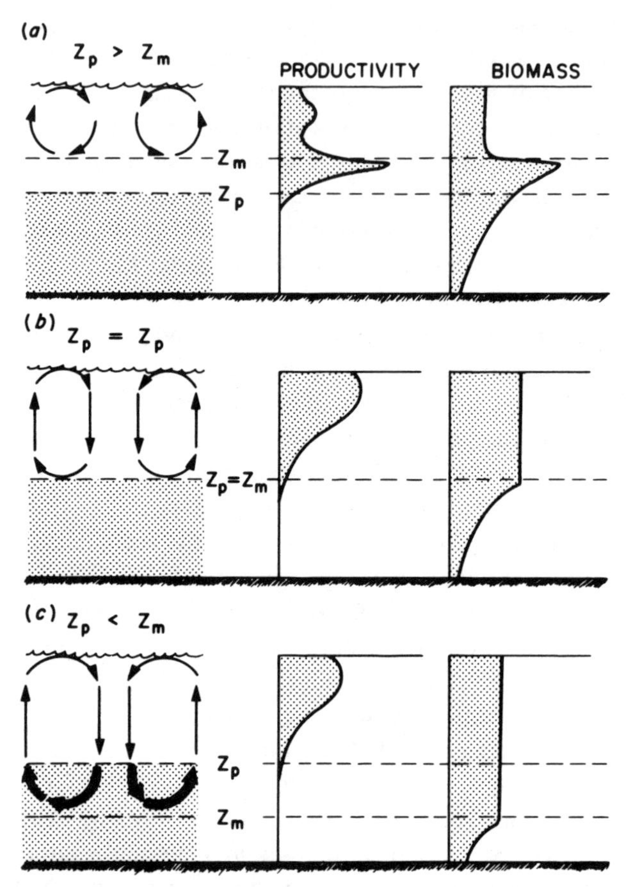

Figure 6.4 Influence of photic-layer depth (Z_p) and mixed-layer depth (Z_m) on the vertical distribution of phytoplankton productivity and biomass. Such vertical patterns help explain differences in phytoplankton productivity and biomass between locations within a single impoundment, between individual reservoirs, and among various types of reservoirs.

algae are exposed not only to rapidly fluctuating light but intermittently to aphotic conditions (Figure 6.4*c*). Because phytoplankton $P < R$ in the aphotic zone, integral primary productivity must be limited to some extent where $Z_p < Z_m$. The degree of light limitation will depend on the relative amount of time cells spend at depths $> Z_p$ and the frequency of their recirculation into the photic zone. Comparisons of the vertical distributions of phytoplankton productivity, biomass, and photochemical capacity in a deeply mixed eutrophic reservoir (Lake Texoma, Oklahoma–Texas) showed that although areal phytoplankton productivity was light-limited, the physiological status of the phytoplankton asemblage was not impaired by intermittent mixing of algal cells into the aphotic zone (Kimmel and White 1979).

The photic layer — mixed layer relationship is also important for understanding the distribution and magnitude of phytoplankton productivity and biomass within individual reservoirs. All three patterns described above and illustrated in Figure 6.4 can occur within a single reservoir as photic layer depth increases and nutrient availability decreases along the riverine-to-lacustrine zone gradient. However, because the photic layer depth increases with improving water clarity, areal primary productivity (mg C m^{-2}) may remain relatively constant along the reservoir longitudinal axis. Thus, traveling downstream in an idealized reservoir one might find 1) light-limited productivity in the mixed layer and a homogeneous vertical distribution of algal biomass in the well-mixed riverine zone ($Z_p < Z_m$, Figure 6.4*c*), 2) somewhat higher volumetric and areal productivity in the mixed layer and a less homogeneous vertical distribution of algal biomass in the transition zone ($Z_p = Z_m$, Figure 6.4*b*), and 3) nutrient-limited productivity in the mixed layer and algal productivity and biomass peaks within the metalimnion of the lacustrine zone ($Z_p > Z_m$, Figure 6.4a).

PHYTOPLANKTON PHOTOINHIBITION

The planktonic environment undergoes rapid fluctuations in light intensity to which phytoplankton respond on temporal scales ranging from seconds (e.g., light-state transitions) to minutes (e.g., chloroplast contraction) to days (e.g., cell doubling time) (Harris 1978, 1980a, Vincent 1979, 1980a, Walsh and Legendre 1983, Falkowski 1984, Elser and Kimmel 1985a). Phytoplankton ecologists commonly observe depressed photosynthesis rates in near-surface samples during fixed-depth in situ productivity measurements (e.g., Goldman et al. 1963, Stadlemann et al. 1974, Ganf 1975, Smith et al. 1980, Platt et al. 1980). The extent of photoinhibition (the suppression of photosynthesis by supraoptimal irradiance) of phytoplankton productivity depends on the intensity, duration, and quality of the exposure irradiance as well as on the light history of the algal asemblage (Harris 1978, Smith et al. 1980). However, phytoplankton cells in well-mixed surface layers are circulated through a light gradient by vertical water movements rather than remaining at fixed light intensities. Therefore, it has been hypothesized that the photoinhibition observed in fixed-depth, static-bottle incubations is a methodological artifact, the result of exposing algal cells to high light intensities for longer time periods (i.e., hours vs. minutes) than normally occur in surface mixed layers (Harris and Piccinin 1977, Gallegos et al. 1978, Harris 1978, 1980a,b, Marra 1978a,b, Gallegos and Platt 1982).

Elser and Kimmel (1985b) employed the chlorophyll in vivo fluorescence (IVF) responses of phytoplankton to evaluate the occurrence of in situ photoinhibition of photosynthesis in several southeastern U.S. reservoirs and used both IVF and ^{14}C-uptake measurements to characterize the responses of phytoplankton assemblages to experimental manipulations of light intensity. Their field results, which were obtained independently of static-bottle techniques, show that phytoplankton photoinhibition does occur in near-surface waters during periods of high incident light and weak vertical mixing. Their experimental results indicate that 1) the severity of photoinhibition and the extent of recovery from photoinhibition depend on the duration of exposure to a high-irradiance environment, and 2) photochemical responses to bright light operate on time scales that, except in very calm conditions or in very shallow mixed layers, should allow vertical mixing to limit the occurrence of in situ photoinhibition by minimizing prolonged exposures of phytoplankton cells to supraoptimal light intensities. Therefore, although these results indicate that photoinhibition of near-surface phytoplankton photosynthesis does occur, they also support the hypothesis that the photoinhibition of phytoplankton productivity occurs less commonly than is suggested by fixed-depth productivity measurements.

HETEROTROPHY BY RESERVOIR PHYTOPLANKTON

In regions with erodible soils (e.g., the midwestern U.S.), turbidity resulting from high concentrations of suspended silts and clays in the water column limits light availability for phytoplankton photosynthesis in many reservoirs (Ellis 1936, Marzolf 1984). Conceivably, the heterotrophic assimilation of dissolved organic compounds could contribute significantly to the carbon balance of light-starved planktonic algae in turbid reservoirs and in other "optically deep" systems (i.e., where $Z_m \gg Z_p$; Talling

1971). Experiments on natural phytoplankton assemblages in a variety of oceanic and lacustrine environments have demonstrated a widespread capability of algae for the assimilation of organic substrates (Parsons and Strickland 1962, Wright and Hobbie 1966, Allen 1969, 1971, McKinley and Wetzel 1979, Vincent and Goldman 1980, Vincent 1980b). Because phytoplankton cells in well-mixed turbid reservoirs spend considerable time in aphotic conditions, algae capable of heterotrophy could have a distinct competitive advantage over obligate photoautotrophs.

White (1981) and Ellis and Stanford (1982) have investigated the occurrence and ecological significance of algal heterotrophy in turbid, mesotrophic Lake Texoma, Oklahoma–Texas. Comparison of autotrophic, photoheterotrophic, and chemoheterotrophic activities of algal cells by microautoradiography revealed that algal heterotrophy occurred primarily in uncommon algae that were not particularly active autotrophs. White concluded that although some algae assimilated organic substrates, 1) neither photoheterotrophy nor chemoheterotropy were major sources of carbon for any phytoplankton species in the Lake Texoma assemblage, and 2) heterotrophic carbon uptake by phytoplankton was negligible relative to photoautotrophy. Similar conclusions were reached by Ellis and Stanford (1982) for phytoplankton in both turbid Lake Texoma and in clear, oligotrophic Flathead Lake, Montana.

In Lake Texoma algal heterotrophy appears to permit the persistence of phytoplankton species that are unable to compete effectively as autotrophs rather than aiding species in becoming dominant. These results 1) support Harris's (1978) contention that the primary importance of algal heterotrophy is that the probability of survival for a phytoplankton cell can be enhanced by a small change in the cellular carbon balance, and 2) suggest that algal heterotrophy may promote algal species diversity and thus provide a partial explanation for the "paradox of the plankton" (Hutchinson 1961, Richerson et al. 1970).

TURBID INFLOWS AND RESERVOIR PHYTOPLANKTON PRODUCTION

The influence of river inflow on lacustrine systems has often been investigated (e.g., Hutchinson 1957), yet few studies have addressed directly the interacting effects of river-borne nutrients and suspended particles on phytoplankton productivity. Because of the large drainage basins of most reservoirs (relative to impoundment surface area and volume), limnological processes in reservoirs are closely coupled to both seasonal and episodic watershed runoff. The movement of water-borne suspended sediments and nutrients resulting from storm-related runoff often occurs as a "plug flow," and variables such as turbidity, nutrient concentrations, and phytoplankton biomass and productivity monitored at any one point in a reservoir show erratic temporal patterns.

Kimmel (1981) examined the effects of episodic turbid inflows on planktonic productivity in Lake Texoma and concluded that the phytoplankton responded to turbid watershed inflow pulses in four phases:

1. Phytoplankton photosynthesis was initially limited by high abiogenic turbidity.
2. Horizontal and vertical displacement of phytoplankton occurred by advection and coflocculation with sedimenting silt and clay aggregates. Algal biomass removal by coflocculation occurred only in association with high concentrations of suspended silts and clays (also see Avnimelech et al. 1982).
3. Nutrient stimulation of phytoplankton photosynthesis occurred when turbidity was reduced to the extent that light availability was no longer the most limiting factor.
4. Phytoplankton productivity returned to its previous level as the advected nutrient supply decreased, and nutrient availability again became the primary limiting factor.

Although occurring on a temporal rather than a spatial scale, this response sequence is quite similar to that acting along the longitudinal axis of a reservoir (e.g., Figure 6.1).

In reservoirs that receive drainage from large erodible watersheds, storm-related inflow events may produce major ecological consequences by virtue of reduced light availability, increased nutrient inputs, and rapid longitudinal and vertical displacement of planktonic organisms and suspended particles, resulting in the episodic "resetting" of physical-chemical conditions and planktonic community structure (e.g., Ellis 1936, Tilzer et al. 1976, Stanford 1978, Goldman and Kimmel 1978, Kimmel 1981). Food-web transfer efficiencies, and thus overall system productivity, may be affected if the plankton community size structure or patterns of seasonal succession are significantly altered.

EMPIRICAL MODELS AND RESERVOIR PRODUCTIVITY

Nutrient-Loading Models and Trophic-State Indices

The cultural eutrophication of lakes and reservoirs has been of concern to citizen and scientist alike for over three decades (e.g., see National Academy of Sciences 1969, Likens 1972), and research to assess the trophic status of lacustrine systems has been conducted internationally (Vollenweider and Kerekes 1980). Consequently, numerous trophic-state indices and nutrient loading — trophic response models have been developed and are now in common use (reviewed by Reckhow 1979a,b, Kennedy and Walker, Chapter 3, this volume). However, the appropriate use of such empirical models and indices for management purposes requires careful consideration of their underlying assumptions, the limitations of the data sets upon which they are based, and the degree of uncertainty associated with their predictions (Reckhow 1979a,b).

For a number of reasons empirical nutrient-loading models and trophic-state indices often yield equivocal results when applied to reservoirs (Lind 1979, Placke and Poppe 1980, Hannan et al. 1981, Higgins et al. 1981, Gloss et al. 1981. Placke 1983):

1. Many reservoirs are located in geographic regions that are poorly represented in most eutrophication model data sets (Thornton 1984; Thornton Chapter 1, this volume), which tend to be dominated by north-temperature-zone natural lakes (e.g., see Lee et al. 1978).
2. Phosphorus loading is not necessarily the primary factor limiting algal growth in reservoirs, as is assumed in most nutrient-loading models, but is only one of many factors that influence algal abundance. In particular, low light availability often moderates the effects of nutrient loading in turbid reservoirs (Kimmel and Lind 1972, O'Brien 1975) and in well-mixed mainstem impoundments (Placke 1983), and water residence time is a critical factor influencing the algal growth response to available nutrients (Soballe and Kimmel 1987).
3. In reservoirs receiving high concentrations of suspended solids, a significant fraction of the total phosphorus loading may be either biologically unavailable (Sonzogni et al. 1982) or rapidly lost to the sediments (Chapra 1980, Gloss et al. 1981). However, this difference may be corrected by application of more suitable phosphorus

sedimentation coefficients (e.g., Jones and Bachmann 1978, Canfield and Bachmann 1981, Higgins et al. 1981).

4. The estimated rate of nutrient loading to a reservoir may bear little relation to the actual rate of nutrient supply to the phytoplankton during the growing season due to the common occurence of density flows (e.g., Ford, Chapter 2, Kennedy and Walker, Chapter 3, Hannan and Cole, Chapter 4).
5. Nutrient retention in shallow, rapidly flushed systems (e.g., mainstem reservoirs) is relatively low as compared to deep lakes with long water residence times. Chapra (1975) and Reckhow (1979) have noted that different mechanisms may govern phosphorus loading–trophic response relationships at overflow rates (z/τ = mean depth in meters/hydraulic residence time in years) greater than 50 m yr^{-1}.

Placke (1983) conducted a trophic status evaluation of the major Tennessee Valley Authority (TVA) tributary storage and mainstem impoundments. As compared with Vollenweider's (1975) data (primarily from north-temperature-zone natural lakes), TVA reservoirs receive high phosphorus loadings but also have high overflow rates ($z/\tau > 50$ m yr^{-1} for 13 of the 18 reservoirs) (Figure 6.5). Placke concluded that the major differences in factors controlling algal productivity and responses to nutrient enrichment in tributary storage vs. mainstem impoundments (e.g., light vs. nutrient limitation, flushing rates, photic layer — mixed layer depth relationships) precluded the valid application of any single trophic-state index to all TVA reservoirs.

The Morphoedaphic Index

The morphoedaphic index (MEI = total dissolved solids in mg L^{-1}/mean depth in m) is another empirical relationship that has been applied widely to predict lake and reservoir productivity (e.g., Oglesby 1982, Ryder 1982, Jenkins 1982). It is particularly useful to fishery managers, as it permits a priori predictions of potential fish production from readily available data on lake morphometry and water chemistry.

Although the MEI has proven to be a robust predictor of fisheries production and yield in a variety of freshwater systems, including reservoirs (Ryder 1965, 1982, Jenkins 1967, 1977, 1982), its ecological basis remains unclear (Oglesby 1977, 1982, Jenkins 1982, Adams et al. 1983).

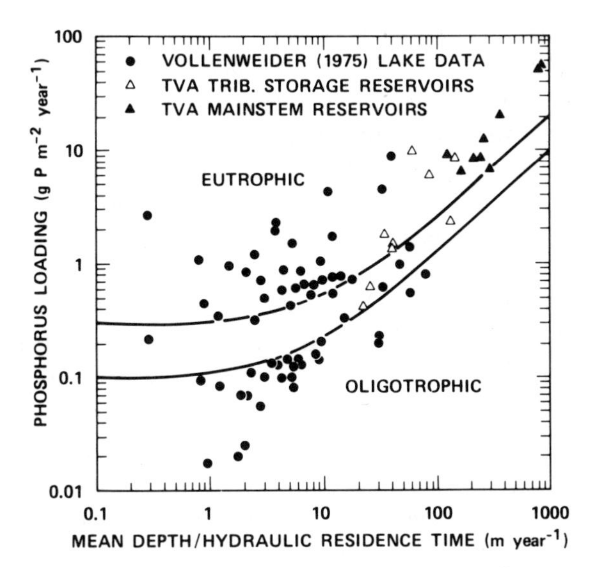

Figure 6.5 Vollenweider's (1968, 1975) phosphorus loading — trophic state relationship applied to north-temperate-zone natural lakes and to 18 Tennessee Valley Authority (TVA) tributary storage and mainstem reservoirs. Values for the more lakelike tributary storage reservoirs lie closer to the natural lake data set than do values for the more riverlike mainstem reservoirs. Modified from Poppe et al. (1980).

The MEI is thought to be indicative of both the availability of nutrients, as reflected by the total dissolved solids concentrations, and the capacity of a system to process energy and materials, as reflected by the mean depth and flushing rate (Oglesby 1977, 1982, O'Brien, Chapter 8, this volume). Presumably, morphological and chemical factors (acting through light, temperature, water residence time, mixing depth, and nutrient availability) control the autochthonous primary production, which by providing the organic matter base of the foodweb determines the magnitude of fish production (Kerr and Martin 1970, Henderson et al. 1973, Ryder et al. 1974, Ryder 1982, Oglesby 1982).

The presumed linkages between the MEI, phytoplankton production, and fish production were examined in an analysis of data for 17 southeastern and midwestern reservoirs (Adams et al. 1983). A significant correlation was found to exist between the MEI and fish production, as previously reported for larger data sets (e.g., Ryder 1982, Jenkins

1982). Phytoplankton productivity and fish production were also significantly correlated, as found for natural lakes and ponds (Melack 1976, Oglesby 1977, Liang et al. 1981), thus supporting the presumed food-chain linkage between phytoplankton and fish production. However, the MEI was not correlated with phytoplankton production, and therefore, the inferred assumption underlying the MEI as a predictor of fisheries yield (i.e., that fish production is proportional to autochthonous primary production as determined by lake basin morphology and water chemistry) was not supported by these data.

The lack of a simple, direct relationship between the MEI and phytoplankton production should not be particularly surprising. Certainly, phytoplankton productivity in lakes and reservoirs is related to morphological and chemical factors, as witnessed by the preceding discussion and an extensive literature on phytoplankton ecology and lake productivity. However, as a system-level descriptor (Kerr 1982), the MEI probably also reflects allochthonous organic matter loading and numerous internal factors, processes, and feedback mechanisms (Ryder 1982, Oglesby 1982) that integrate in such a way as to influence fish production more significantly, or perhaps just more consistently, than phytoplankton production. Although deceptively simple in form, the MEI and similar composite indices represent complex aggregations of covarying physical, chemical, and biological factors, and their ecological foundations will remain vague until we improve our knowledge of specific processes and interactions.

PHYTOPLANKTON PRODUCTIVITY IN MULTIPLE-RESERVOIR SERIES

Dam construction associated with water and power development projects has converted rivers throughout the world into regulated series of man-made impoundments. The effects of reservoir releases on downstream water temperature, flow regimes, water quality, and biota have received considerable attention (Young et al. 1972, Ridley and Steel 1975, Ward 1976, Ward and Stanford 1979, 1981, Petts 1984). However, little is known of the ecological and limnological interactions occurring within impoundment series. While natural lakes can also occur in series (e.g., pater noster lakes), the hypolimnetic releases of water and nutrients from deep-discharge reservoirs maximize the potential influence of upstream impoundments on the physical, chemical, and biological processes occurring in reservoirs downstream (Neel 1963, Wright 1967, Stroud and Martin 1973, Martin and Arneson 1978, Elser and Kimmel 1985b).

The limnological studies of the Vltava Cascade reservoir series by the Hydrobiological Laboratory of the Czechoslovak Academy of Sciences represent the most comprehensive examination of a multiple-reservoir system yet undertaken (Hrbacek and Straskraba 1966, 1973a, 1973b). These studies of two large mainstem storage impoundments (Slapy and Orlik Reservoirs) and three small re-regulation impoundments clearly demonstrate the important influence of upstream reservoir releases on downstream systems and the ecological significance of water-residence time (Straskraba et al. 1973, Javornicky and Komarkova 1973, Straskraba and Javornicky 1973).

On an annual basis impoundments are efficient sediment and nutrient traps (e.g., Dendy et al. 1973, Heinemann et al. 1973, Gloss et al. 1980, 1981), and thus, upstream impoundments often enhance the predominance of lacustrine conditions in reservoirs lying downstream (e.g., Paulson et al. 1980, Paulson 1981, Van Den Avyle et al. 1982, Elser and Kimmel 1985a). Except in new reservoirs during the initial period of basin filling and nutrient leaching from inundated soils (Baxter 1977, Ostrofsky and Duthie 1980, Grimard and Jones 1982, Kimmel and Groeger 1985), the presence of a reservoir reduces the annual nutrient loading to downstream aquatic systems as a result of sedimentation losses of nutrients in the reservoir basin (Gloss et al. 1981, Paulson and Baker 1981, Daley et al. 1981). However, the presence of deep-discharge impoundments upstream also changes the seasonal patterns of hydrologic and nutrient inputs to downstream reservoirs.

Lake Mead (Arizona–Nevada)

Changes in the productivity of Lake Mead resulting from the impoundment of Lake Powell on the Colorado River demonstrate the profound ecological effects of upstream impoundments on reservoirs located downstream. The Colorado River was unregulated prior to the impoundment of Lake Mead by Hoover Dam in 1935. Hoover Dam stabilized flows and reduced suspended sediment loads downstream, but Lake Mead continued to receive silt-laden inflows from the upper Colorado River basin. The Colorado River contributed 97% of the suspended inputs to Lake Mead, and up to 140×10^6 Mg (= metric tons) of suspended sediment entered the reservoir in years of high runoff.

The limnology and productivity of Lake Mead were strongly influenced

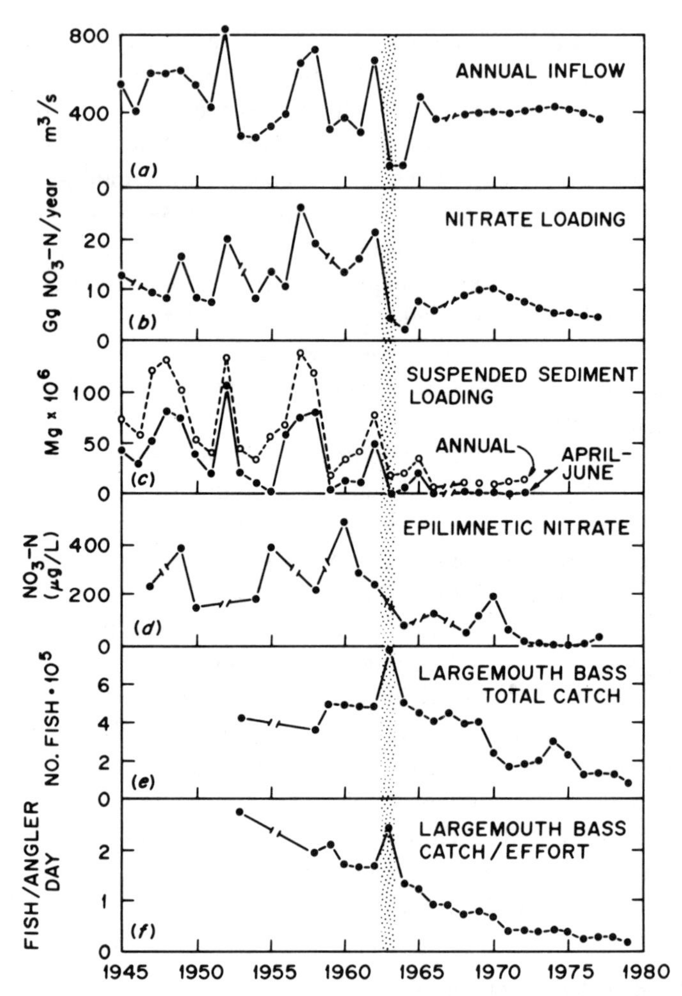

Figure 6.6 Effects of an upstream impoundment (Lake Powell) on the limnology and ecology of Lake Mead (Arizona–Nevada). The shaded bar indicates the date of completion of the Glen Canyon Dam and the impoundment of Lake Powell. Modified from Paulson and Baker (1981) and Prentki and Paulson (1983).

by nutrient-rich turbid overflows that occurred during the spring and early summer period of high Colorado River inflow (Anderson and Pritchard 1951, Hoffman and Jonez 1973) prior to construction of the Glen Canyon Dam 450 km upstream and the formation of Lake Powell in 1963 (Figure 6.6). Lake Powell now functions as a sink for suspended sediments and associated nutrients (Evans and Paulson 1983), and the water discharged from the Glen Canyon Dam to flow downstream to Lake Mead is colder, clearer, and lower in nutrients than prior to 1963. Lake Powell retains 70% of the dissolved phosphorus and 96% of the total phosphorus that once flowed into Lake Mead (Gloss et al. 1980, 1981). Most of the Colorado River inflow to Lake Mead now enters as a hypolimnetic interflow, and consequently, nutrient inputs to the photic layer have been significantly reduced (Paulson and Baker 1981, Prentki and Paulson 1983). Phytoplankton production is now more dependent on internal nutrient recycling than on nutrient advection, and Lake Mead productivity, on a reservoir-wide basis, has declined significantly since 1963.

While the upper basin of Lake Mead has become less productive and more phosphorus-limited, the lower basin has become more productive and nitrogen-limited. Phosphorus-rich wastewater discharges from the city of Las Vegas to the downstream portion of the reservoir (specifically, to Las Vegas Bay and Boulder Basin; see Paulson and Baker 1981 or Prentki and Paulson 1983) have increased steadily since 1963. The morphology and hydrodynamics of Lake Mead are such that this additional phosphorus load is confined to the lower basin, where it enhances phytoplankton production, decreases nitrate availability, and has resulted in the development of nitrogen limitation (Paulson 1981). This unusual situation, which provides a marked contrast to the typical longitudinal pattern of reservoir phytoplankton productivity, is the result of a reversal of the "polarity" of phosphorus loading to Lake Mead since 1963 (i.e., decreased loading to the upper basin due to the formation of Lake Powell and increased loading to the lower basin due to municipal wastewater discharges).

Lake Mead has historically supported an excellent largemouth bass fishery (Hoffman and Jonez 1973). However, the total catch has decreased from about 800,000 fish in 1963 to 105,000 in 1979 despite increased angler pressure (Figure 6.6). The causes of this decline have not been conclusively established. However, the reduced fishery yield is undoubedly related to an overall decline in the biological productivity of Lake Mead since the formation of Lake Powell (Paulson et al. 1980, Baker and Paulson 1983).

Kootenay Lake (British Columbia)

A three-year multidisciplinary study of Kootenay Lake, British Columbia, produced similar conclusions concerning the influence of impoundments on the biological productivity of lacustrine systems located downstream (Daley et al. 1981). Kootenay Lake, an intermontane, fjord-like natural lake, receives 75% of its annual inflow from the Kootenai, Duncan and Lardeau rivers. The operation of dams constructed on the Duncan River (Duncan Dam completed in 1967) and on the Kootenai River (Libby Dam completed in 1973) has altered the natural pattern of water inflow to Kootenay Lake with consequent effects on lake productivity.

The primary effect of the dams was the reduction of the nutrient supply to the Kootenay Lake phytoplankton as mediated by 1) nutrient retention in the reservoirs, 2) enhancement of nutrient uptake by riverine periphyton below the dams, and 3) a shift in nutrient loadings from spring and summer to winter (Figure 6.7). Daley et al. (1981) estimated that 25–50% of the decrease in Kootenay Lake biological productivity was due to the presence of the upstream impoundments and predicted that levels of phytoplankton and zooplankton would be reduced eventually to less than half of what they would have been had the dams not been constructed.

The declining productivities of Lake Mead and Kootenay Lake have produced regional concerns for the future of their sport fisheries. Consequently, methods of enhancing fisheries production by increasing the nutrient availability for phytoplankton production for these systems are now being considered after years of water-quality-related efforts to reduce the nutrient loading. In this regard reservoirs have more options available for productivity-oriented management than do natural lakes (e.g., epilimnetic, hypolimnetic, or selected-depth discharge). Paulson has argued that expensive tertiary treatent of sewage prior to its discharge into Lake Mead is illogical if recovery of the depressed Lake Mead fishery is considered to be a high priority. More recently, Paulson et al. (unpublished data) have demonstrated the importance of increasing nutrient availability in Lake Mead for improving the reservoir fishery by conducting a large-scale experimental fertilization that resulted in a significant enhancement of fishery yield in one tributary arm of the reservoir.

Numerous authors have emphasized the importance of outlet depth in regulating reservoir nutrient balance (Neel 1963, Wright 1967, Stroud and Martin 1973, Soltero and Wright 1975, Baxter 1977, Paulson and Baker 1981, Priscu et al. 1982). Wright (1967) hypothesized that reservoirs with hypolimnetic discharges store heat and dissipate nutrients during the growing season as a result of releasing cold, nutrient-rich water, whereas surface-outflow reservoirs and natural lakes dissipate heat and trap nut-

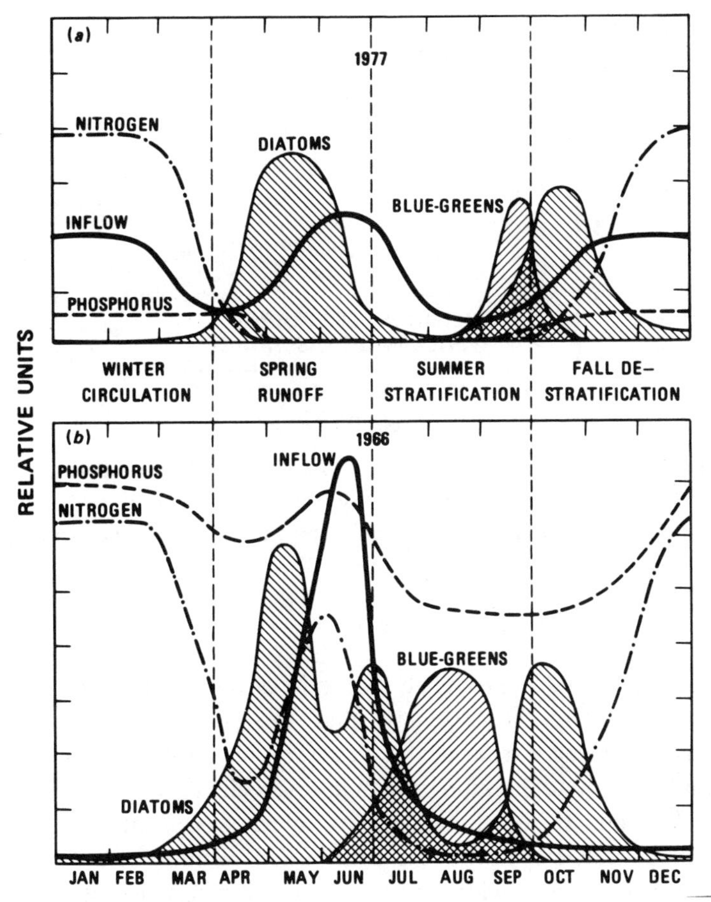

Figure 6.7 Schematic representation of seasonal patterns of streamflow, nutrient concentrations, and phytoplankton biomass in Kootenay Lake, British Columbia: (*a*) in 1977, after contruction of dams on two major tributaries to the lake, and (*b*) in 1966, prior to construction of the dams. Modified from Daley et al. (1981).

rients by releasing warm, nutrient-poor water. Martin and Arneson's (1978) comparison of heat and nutrient distributions in a deep-discharge reservoir and in a surface-discharge lake supported Wright's hypothesis, but the in-lake biological impacts of the differing discharge regimes were difficult to discern. Paulson (1981) constructed a nutrient retention-loss model for Lake Mead to illustrate that epilimnetic rather than hypolimnetic releases of water from Hoover Dam would enhance the productivity of

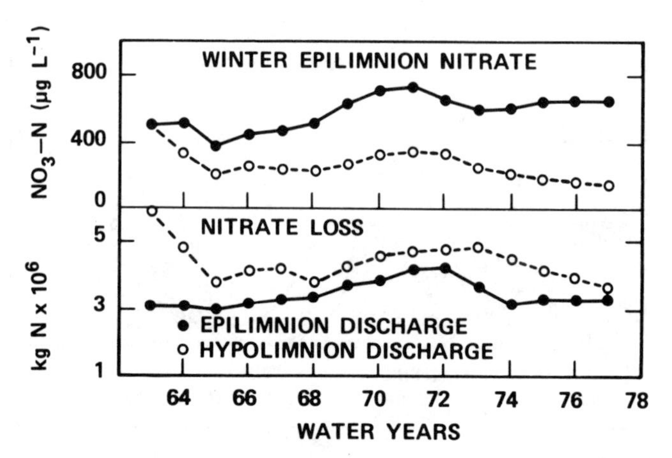

Figure 6.8 Hypothesized differences in Lake Mead (Arizona–Nevada) epilimnetic nitrate concentrations and in NO_3-N losses via discharge from Hoover Dam under conditions of epilimnetic and hypolimnetic releases. Modified from Paulson (1981).

the reservoir by reducing NO_3-N losses from the reservoir (Figure 6.8). Wright's (1967) hypothesis suggests that, in series of deep-discharge impoundments, releases from upstream reservoirs should increase nutrient availability for phytoplankton growth in reservoirs downstream during the growing season. If so, the effects of interreservoir nutrient transfers should be reflected in the spatial distribution of nutrient availability for phytoplankton growth within and among the reservoirs of a multiple-impoundment series.

Elser and Kimmel (1985b) investigated the availability of nitrogen (N) and phosphorus (P) for phytoplankton production and the importance of interreservoir nutrient transfers in a headwater-to-mainstem series of three hypolimnetic-discharge impoundments (Norris, Melton Hill, and Watts Bar reservoirs) in eastern Tennessee. Dissolved nutrient analyses and enrichment experiments indicated that P, rather than N, was the primary limiting nutrient for phytoplankton growth throughout the reservoir series. Phytoplankton P deficiency, as reflected by chlorophyll-specific alkaline phosphatase activity (APA), peaked in late August and then decreased during the fall and winter. Late in the growing season APA levels increased downstream within each reservoir, reflecting uplake-to-downlake decreases in P availability within the reservoir, but decreased downstream from reservoir to reservoir, indicating increased P availability down the impoundment series.

In the Norris — Melton Hill — Watts Bar reservoir series hypolimnetic releases of water and nutrients from upstream impoundments increased nutrient availability during the growing season (Elser and Kimmel 1985b) and thus enhanced phytoplankton production in impoundments downstream. Similar within-reservoir and among-reservoir patterns in nutrient availability and phytoplankton productivity have been observed in the Hartwell — R. B. Russell — Clarks Hill reservoir series on the Savannah River, Georgia-South Carolina (Kimmel et al. 1988).

In the absence of an upstream impoundment nutrient inputs to reservoirs are usually seasonal and are associated primarily with periods of high precipitation or snowmelt in winter and spring (Daley et al. 1981, Ward and Stanford 1983). However, because of suboptimal growth conditions (low light, low water temperature, high turbidity) and high flushing rates during such times of maximal inflow, nutrient inputs may not be utilized as efficiently by reservoir phytoplankton as nutrient inputs occurring during the growing season. Although the total nutrient flux downstream is reduced, the retention of spring runoff in storage reservoirs and the subsequent gradual discharge of hypolimnetic water and nutrients as a consequence of hydropower operations potentially extends the period of downstream nutrient transport through the growing season. Elser and Kimmel (1985b) hypothesized that this temporal redistribution of the nutrient supply to downstream reservoirs results in a more efficient conversion of the available nutrients to phytoplankton production than would occur in the absence of upstream, deep-discharge impoundments.

This hypothesis presumes that a significant portion of the nutrients released from an upstream impoundment during the growing season enters the euphotic zone of the reservoir downstream (e.g., via mechanisms such as those described by Carmack and Gray 1982 and Fischer and Smith 1983) rather than being confined to aphotic layers (as now occurs in Lake Mead). Density flows are often enhanced by hypolimnetic releases, however, and the effective nutrient supply to the trophogenic zone of a thermally stratified downstream reservoir may be significantly less than is indicated by the riverine nutrient loading to the reservoir (Paulson and Baker 1981). An improved understanding of the extent to which nutrients released in hypolimnetic discharges enter or are isolated from the euphotic layer in impoundments downstream is required to determine more specifically the influence of deep-discharge releases on the biological productivity of multiple-impoundment series.

Series of large, multiple-purpose reservoirs (e.g., those located on the Columbia, Colorado, Missouri, Arkansas, and Tennessee Rivers) are operated on an integrated river-basin scale to provide flood control, hydroelectric power, water supply, and in some cases navigation. These results suggest that during the growing season reservoir hypolimnetic releases associated with hydropower operations influence the supply of nutrients available for phytoplankton production in reservoir series, a result consistent with the implication of Wright's (1967) hypothesis regarding the downstream dissipation of nutrients by deep-discharge reservoirs. The studies discussed above also clearly indicate that the productivity and ecology of reservoirs located in multiple-impoundment series must be viewed in the context of the limnological dynamics resulting from basin-level reservoir operations.

It remains difficult to evaluate specifically the importance of inter-reservoir transfers because other factors (e.g., differences in drainage basin size, reservoir basin morphology, hydrodynamics, intermediate nutrient sources, and limiting factors other than nutrients) also influence nutrient availability and phytoplankton productivity in reservoir series. However, because of the potential value of integrating multiple-impoundment operational and scheduling practices with reservoir fishery production and water quality management objectives, further research to determine the influence of interreservoir nutrient transfers on biological productivity in multiple-impoundment series is warranted.

CONCLUSION

We began this chapter by suggesting that a conceptual view of reservoirs as "river-lake hybrids" is useful for interpreting the spatial heterogeneity and limnological differences commonly observed within and among reservoir systems. Kerr (1982) concluded that "internal" (i.e., reductionist) methods of ecosystem analysis are unlikely to succeed unless preceded by relevant observation and theory at the "external" (holistic) level. The river-lake hybrid analogy is holistic in the same sense as Kerr's "external" analysis and, similarly, tends to point to specific processes and relationships that require further, more detailed investigation (i.e., internal analysis). Obviously, both "external" and "internal" levels of analysis are required to advance our knowledge of reservoirs as ecological systems.

The close coupling of man-made impoundments to their watersheds and their semifluvial nature make reservoirs particularly heterogeneous systems. While their spatial and temporal variability makes reservoirs difficult to categorize, it affords both challenges and opportunities to ecologists and limnologists. The diversity of environmental conditions existing within and among impoundments provides unique opportunities for addressing a variety of research questions. For example, reservoir plankton communities appear to offer excellent opportunities for the investigation of ecological responses to episodic environmental disturbances and ecosystem resilience. The longitudinal gradients in physical and chemical factors that exist in many reservoirs provide opportunities to address questions regarding ecological, food web, and nutrient cycling relationships occurring in different environments (e.g., in riverine vs. lacustrine conditions) but within the same physical system. Additionally, the fact that reservoirs can be (and are) manipulated provides high potential for both planned and "opportunistic" experimentation. If we are clever enough to use them to our advantage, the peculiar features of reservoirs (e.g., their river-lake hybrid nature, hypolimnetic outlets, fluctuating water levels and volumes) can permit unique insights into the structure and function of both lotic and lentic ecosystems.

ACKNOWLEDGMENTS

Numerous persons have made direct and indirect contributions to the content of this chapter. We thank the other contributors to this volume, especially H. H. Hannan, R. H. Kennedy, and K. W. Thornton, for stimulating discussion, critical comment, and helpful suggestions. We also thank C. C. Coutant, C. W. Gehrs, A. W. Groeger, D. M. Soballe, J. A. Solomon, and W. Van Winkle for reviewing the manuscript.

Portions of the research discussed in this paper were sponsored by the Office of Water Research and Technology, U.S. Department of the Interior, through the Water Resources Research Institutes of the states of Oklahoma (B.L.K.), Texas (O.T.L.), and Nevada (L.J.P.); by the U.S. Environmental Protection Agency and the U.S. Bureau of Reclamation (L.J.P.); and by the University of Oklahoma, the Environmental Laboratory, U.S. Army Waterways Experiment Station, and the Office of

Health and Environmental Research, U.S. Department of Energy (B.L.K.).

Oak Ridge National Laboratory is operated by Martin-Marietta Energy Systems, Inc. under Contract No. DE-AC05-840R21400 with the U.S. Department of Energy. Publication No. 3185. Environmental Sciences Division, Oak Ridge National Laboratory.

REFERENCES

Abernathy, A. R. and H. R. Bungay. 1972. Water quality predications based on limnology parameters. Water Resources Research Institute, Report No. PB 22004, Clemson University, Clemson. SC.

Ackermann, W. C., G. F. White, and E. B. Worthington, eds. 1973. Man-made lakes: Their problems and environmental effects. Geophys. Monogr. 17, Am. Geophys. Union, Washington, DC, 847 pp.

Adams, S. M., B. L. Kimmel, and G. R. Ploskey. 1983. Organic matter sources for reservoir fish production: A trophic-dynamics analysis. *Can. J. Fish. Aq. Sci.* 40:1480–1495.

Allen, H. L. 1969. Chemo-organotrophic utilization of dissolved organic compounds by planktonic algae and bacteria in a pond. *Int. Rev. Ges. Hydrobiol.* 54:1–33.

Allen, H. L. 1971. Dissolved organic carbon utilization in size-fractionated algal and bacterial communities. *Int. Rev. Ges. Hydrobiol.* 56:731–749.

Anderson, J. L. and G. L. Hergenrader. 1973. Comparative preliminary productivity of two flood control reservoirs in the Salt Valley watershed of eastern Nebraska. *Trans. Nebr. Acad. Sci.* 2:134–143.

Anderson, E. R. and D. W. Pritchard. 1951. Physical limnology of Lake Mead. Rept. No. 258, U.S. Navy Electronics Laboratory, San Diego, CA. 153 pp.

Anderson, G. C. 1969. Subsurface chlorophyll maximum in the Northeast Pacific Ocean. *Limnol. Oceanogr.* 14:386–391.

Avnimelech, Y., B. W. Troeger, and L. W. Reed. 1982. Mutual flocculation of algae and clay: Evidence and implications. *Sci.* 216:63–65.

Bacon, E. J. 1973. Primary productivity, water quality, and bottom fauna in Doe Valley Lake, Meade County, Kentucky. Ph.D. dissertation, University of Louisville, Louisville, KY. 183 pp.

Baker, J. R. and L. J. Paulson. 1983. The effects of limited food availability on the striped bass fishery in Lake Mead. Pages 551–561 in V. D. Adams and V. A. LaMarra, eds. Aquatic resources management of the Colorado River ecosystem. Ann Arbor Science, Ann Arbor, MI. 697 pp.

Barko, J. W. 1981. The influence of selected environmental factors on submersed macrophytes — a summary. Pages 1378–1382 in H. G. Stefan, ed. Proc. symposium on surface water impoundments. American Society Civil Engineers, New York, NY. 1682 pp.

Barnes, R. K. and K. H. Mann. 1980. Prologue. Pages 1–3 in R. K. Barnes and K.H. Mann, eds. Fundamentals of aquatic ecosystems. Blackwell Sci., London, England.

Bartell, S. M., A. L. Brenkert, R. V. O'Neill, and R. H. Gardner. 1988. Temporal variation in regulation of production in a pelagic food web model. Pages 101–118 in S. R. Carpenter, ed. Complex interactions in lake communities. Springer-Verlag, New York, NY.

Baxter, R. M. 1977. Environmental effects of dams and impoundments. *Annu. Rev. Ecol. Syst.* 8:255–283.

Beattie, M., H. J. Bromley, M. Chambers, C. Goldspink, J. Vijverberg, N. P. van Zalingen, and H. L. Golterman. 1972. Limnological studies on Tjeukemeer — a typical Dutch "older reservoir." Pages 421–446 in Z. Kajak and A. Hillbricht-Ilkowska, eds Productivity problems of freshwaters. PWN Polish Sci. Publ., Warsaw, Poland.

Benson, N. G. 1982. Some observations on the ecology and fish management of reservoirs in the United States. *Can. J. Water Resours.* 7:2–25.

Berman, T. and U. Pollingher. 1974. Annual and seasonal variations of phytoplankton, chlorophyll and primary production in Lake Kinneret. *Limnol. Oceanogr.* 19:266–276.

Brook, A. J. and W. B. Woodward. 1956. Some observations on the effects of water inflow and outflow on the plankton of small lakes. *J. An. Ecol.* 25:22–35.

Brook, A. J. and J.Rzoska. 1954. The influence of the Gebel Aulia Dam on the development of Nile plankton. *J. Anim. Ecol.* 23:101–115.

Brooks, A. S. and B. G. Torke. 1977. Vertical and seasonal distribution of chlorophyll *a* in Lake Michigan. *J. Fish. Res. Board Can.* 34:2280–2287.

Brylinsky, M. 1980. Estimating the productivity of lakes and reservoirs. Pages 411–453 in E. D. LeCren and R. H. Lowe-McConnell, eds. The functioning of freshwater ecosystems. Cambridge Univ. Press, London, England. 588 pp.

Brylinsky, M. and K. H. Mann. 1973. An analysis of factors governing productivity in lakes and reservoirs. *Limnol. Oceanogr.* 18:1–14.

Canfield, D. E. and R. W. Bachmann. 1981. Prediction of total phosphorus concentrations, chlorophyll *a* and secchi depths in natural and artificial lakes. *Can. J. Fish. Aquat. Sci.* 38:414–423.

Carline, R. F. 1986. Indices as predictors of fish community traits. Pages 46–56 in G. E. Hall and M. J. Van Den Avyle, eds. Reservoir fisheries management: Strategies for the 80's. Am. Fish Soc., Bethesda, MD.

Carmack, E. C., C. B. Gray, C. H. Pharo, and R. J. Daley. 1979. Importance of lake-river interaction on seasonal patterns in the general circulation of Kamloops Lake, British Columbia. *Limnol. Oceanogr.* 24:634–644.

Carmack, E. C. and C. B. J. Gray. 1982. Patterns of circulation and nutrient supply in a medium residence time reservoir, Kootenay Lake, British Columbia. *Can. J. Water Resours.* 7:51–70.

Carpenter, S. R., J. F. Kitchell, J. R. Hodgson, P. A. Cochran, J. J. Elser, M. M. Elser, D. M. Lodge, D. Kretchmer, X. He, and C. N. von Ende. 1987. Regulation of lake primary productivity by food web structure. *Ecol.* 68:1863–1876.

Carpenter, S. R., J. F. Kitchell, and J. R. Hodgson. 1985. Cascading trophic interactions and lake productivity. *Bioscience* 35:634–639.

Carpenter, S. R. and J. F. Kitchell. 1984. Plankton community structure and limnetic primary production. *Am. Nat.* 124:159–172.

Chamberlain, L. L. 1972. Primary productivity in a new and an older California reservoir. *Calif. Fish and Game* 58:254–267.

Chang, W. Y. and D. G. Frey. 1977. Monroe Reservoir, Indiana. Part 2: Nutrient relations. Pages 71–117 in Indiana University Water Resour. Res. Center Tech. Rep. 87.

Chapra, S. C. 1980. Application of phosphorus loading models to river-run lakes and other incompletely mixed systems. Pages 329–334 in Restoration of lakes and inland waters, U.S. EPA 440/5-281-010.

Chapra, S. C. 1975. Comment on "An empirical method of estimating the retention of phosphorus in lakes," by W. B. Kirchner and P. J. Dillon. *Water Resources Res.* 2:1033–1034.

Claflin, T. O. 1968. Reservoir aufwuchs on inundated trees. *Trans. Am. Microsc. Soc.* 87:97–104.

Coutant, C. C. 1963. Stream plankton above and below Green Lane Reservoir. *Proc. Penn. Acad. Sci.* 37:122–126.

Cowell, B. C. 1970. The influence of plankton discharges from an upstream reservoir on standing corps in a Missouri River reservoir. *Limnol. Oceanogr.* 15:427–441.

Cowell, B. C. and P. L. Hudson. 1967. Some environmental factors influencing benthic invertebrates in two Missouri River reservoirs. Pages 541–555 in Reservoir fishery resources. Spec. Publ. Am. Fish Soc., Washington, DC.

Crayton, W. M. and M. R. Sommerfeld. 1981. Impacts of a desert impoundment on the phytoplankton community of the lower Colorado River. Pages 1608–1617 in H. G. Stefan, ed. Proceedings of the symposium on surface water impoundments. Am. Soc. Civil Engr., New York, NY. 1682 pp.

Crumpton, W. G. and R. G. Wetzel. 1982. Effects of differential growth and mortality in the seasonal succession ofphytoplankton populations in Lawrence Lake, Michigan. *Ecology* 63:1729–1739.

Cummins, K. W. 1979. The natural stream ecosystem. Pages 7–24 in J. V. Ward and J. A. Stanford, eds. The ecology of regulated streams. Plenum Press, New York, NY.

Cummins, K. W. 1974. Structure and function of stream ecosystems. *Bioscience* 24:631–641.

Daley, R. J., E. C. Carmack, C. B. J. Gray,C. H. Pharo, S. Jasper, and R. C. Wiegand. 1981. The effects of upstream impoundments on the limnology of Kootenay Lake. B. C. Environ. Con. Sci. Rept. No. 117, Vancouver, B. C. 98 pp.

Davies, W. D., W. L. Shelton, D. R. Bayne, and J. M. Lawrence. 1980. Fisheries and limnological studies on West Point Reservoir, Alabama–Georgia. Final report, Mobile District Corps of Engineers, Mobile, AL. 238 pp.

Dendy, F. E., W. A. Champion, and R. B. Wilson. 1973. Reservoir sedimentation surveys in the United States. Pages 349–358 in W. C. Ackermann, G. F. White, and E. B. Worthington, ed. Man-made lakes: Their problems and environmental effects. Geophys. Monogr. 17, Am. Geophys. Union, Washington, DC.

Dickman, M. 1969. Some effects of lake renewal on phytoplankton productivity and species composition. *Limnol. Oceanogr.* 14:660–666.

Dillon, C. R. and J. H. Rodgers. 1980. Thermal effects on primary productivity of phytoplankton, periphyton and macrophytes in Lake Keowee, S. C. Report No. 81, Water Resources Research Institute, Clemson University, Clemson, SC.

Ducklow, H. W. 1983. Production and fate of bacteria in the oceans. *Bioscience* 33:494–501.

Ellis, B. K. and J. A. Stanford. 1982. Comparative photoheterotrophy, chemoheterotrophy, and photolithotrophy in a eutrophic reservoir and an oligotrophic lake. *Limnol. Oceanogr.* 27:440–454.

Ellis, M. M. 1936. Erosion silt as a factor in aquatic environments. *Ecology* 17:29–42.

Elser, J. J. and B. L. Kimmel. 1985a. Photoinhibition of temperate lake phytoplankton by near-surface irradiance: Evidence from vertical profiles and field experiments. *J. Phycol.* 21:419–427.

Elser, J. J. and B. L. Kimmel. 1985b. Nutrient availability for phytoplankton production in a multiple-impoundment series. *Can. J. Fish. Aquat. Sci.* 42:1359–1370.

Evans, T. D. and L. J. Paulson. 1983. The influence of Lake Powell on the suspended sediment-phosphorus dynamics of the Colorado River inflow to Lake Mead. Pages 57–68 in V. D. Adams and V. A. LaMarra, eds. Aquatic resources management of the Colorado River ecosystem. Ann Arbor Science, Ann Arbor, MI.

Falkowski, P. G. 1984. Physiological responses of phytoplankton to natural light regimes. *J. Plankton Res.* 6:295–307.

Fee, E. J. 1976. The vertical and seasonal distribution of chlorophyll in lakes of the Experimental Lakes Area, northwestern Ontario: Implications for primary production estimates. *Limnol. Oceanogr.* 21:767–783.

Findenegg, I. 1964. Types of planktonic primary production in the lakes of the Eastern Alps as found by the radioactive carbon method. *Verh. Int. Verein. Limnol.* 15:352–359.

Fischer, H. B. and R. D. Smith. 1983. Observations of transport to surface waters from a plunging inflow to Lake Mead. *Limnol. Oceanogr.* 28:258–272.

Fogg, G. E. 1975. Algal cultures and phytoplankton ecology, 2nd ed. University of Wisconsin Press, Madison, WI. 175 pp.

Ford, D. E. and K. W. Thornton. 1979. Time and length scales for the one-dimensional assumption and its relation to ecological models. *Wat. Resours. Res.* 15:113–120.

Fraser, J. E. 1974. Water quality and phytoplankton productivity of Summersville Reservoir. *W. Va. Acad. Sci.* 8:8–16.

Gak, D. Z., V. V. G. Furvich, I. L. Korelyakova, L. E. Kostikova, N. A. Konstantinova, G. A. Olivari, A. D. Primachenko, Y. Y. Tseeb, K. S. Vladimirova, and L. N. Zimbalevskaya. 1972. Productivity of aquatic organism communities of different trophic levels in Kiev Reservoir. Pages 447–456 in A. Kajek and A. Hillbricht-Ilkowska, eds. PW N Polish Sci. Publ., Warsaw, Poland.

Gallegos, C. L., G. M. Hornberger, and M. G. Kelly. 1980. Photosynthesis-light relationships of a mixed culture of phytoplankton in fluctuating light. *Limnol. Oceanogr.* 25:1982–1092.

Gallegos, C. L. and T. Platt. 1982. Phytoplankton production and water motion in surface mixed layers. *Deep Sea Res.* 29:65–76.

Ganf, G. G. 1975. Photosynthetic production and irradiance-photosynthesis relationships of the phytoplankton from a shallow equatorial lake (L. George, Uganda). *Freshwater Biol.* 5:13–39.

Gascon, D. and W. C. Leggett. 1977. Distribution, abundance, and resource utilization of littoral zone fishes in response to a nutrient/production gradient in Lake Memphremagog. *J. Fish. Res. Board Can.* 34:1105–1117.

Gloss, S. P, R. C. Reynolds, L. M. Mayer, and D. E. Kidd. 1981. Reservoir influences on salinity and nutrient fluxes in the arid Colorado River basin. Pages 1618–1629 in H. G. Stefan, ed. Proceedings of the symposium on surface water impoundments. Am. Soc. Civil Engr., New York, NY.

Gloss, S. P, L. M. Mayer, and D. E. Kidd. 1980. Advective control of nutrient dynamics in the epilimnion of a large reservoir. *Limnol. Oceanogr.* 25:219–228.

Goldman, C. R. and B. L. Kimmel. 1978. Biological processes associated with suspended sediment and detritus in lakes and reservoirs. Pages 19–44 in J. Cairns, E. F. Benefield, and J. R. Webster, eds. Current perspectives on river-reservoir ecosystems. N. Am. Bentho. Soc. Publ. 1.

Goldman, C. R. 1968. Aquatic primary production. *Am. Zool.* 8:31–42.

Goldman, C. R., D. T. Mason, and B. J. Wood. 1963. Light injury and inhibition in Antarctic freshwater phytoplankton. *Limnol. Oceanogr.* 8:313–22.

Grimard, Y. and H. G. Jones. 1982. Trophic upsurge in new reservoirs: A model for total phosphorus concentrations. *Can. J. Fish Aquat. Sci.* 39:1473–1483.

Groeger, A. W. and B. L. Kimmel. 1984. Organic matter supply and processing in lakes and reservoirs. Pages 282–285 in Lake and Reservoir Management. EPA 440/5/84-001. U.S. Environmental Protection Agency, Washington, DC.

Groeger, A. W. and B. L. Kimmel. 1988. Photosynthetic carbon metabolism by phytoplankton in a nitrogen-limited reservoir. *Can. J. Fish. Aquat. Sci.* 45:720–730.

Groeger, A. W. 1979. Organic matter flow through Lake Isabella and the Chippewa River. M. S. Thesis, Central Michigan University, Mt. Pleasant, MI. 76 pp.

Hains, J. J. 1987. Preimpoundment and postimpoundment algal flora of the Savannah River at Richard B. Russell reservoir site. Pages 53–64 in R. H. Kennedy, ed. Supplemental Limnological Studies at Richard B. Russell and Clarks Hill Lakes, 1983–1985. Misc. Paper E-87-2, U.S. Army Engineer Waterways Experiment Station, Vicksburg, MS.

Hammerton, D. 1972. The Nile River — a case history. Pages 171–214 in R. T. Oglesby, C. A. Carlson, and J. A. McCann, River Ecology and Man. Academic Press, New York, NY.

Hannan, H. H., D. Barrows, and D. C. Whitenberg. 1981. The trophic status of a deep-storage reservoir in Central Texas. Pages 425–434 in H. G. Stefan, ed. Proceedings of the symposium on surface water impoundments. Am. Soc. Civil Engr., New York, NY.

Harris, G. P. 1980b. The measurement of photosynthesis in natural populations of phytoplankton. Pages 129–187 in I. Morris, ed. The physiological ecology of phytoplankton. University of California Press, Berkeley, CA.

Harris, G. P 1988. Structural aspects of phytoplankton succession. *Verh. Internat. Verein. Limnol.* 23:2221–2225.

Harris, G. P. 1980a. Temporal and spatial scales in phytoplankton ecology; Mechanisms, methods, models, and management. *Can. J. Fish. Aquat. Sci.* 37:877–900.

Harris, G. P and B. B. Piccinin. 1977. Photosynthesis by natural phytoplankton populations. *Arch. Hydrobiol.* 80:405–57.

Harris, G. P. 1978. Photosynthesis, productivity and growth: The physiological ecology of phytoplankton. *Arch. Hydrobiol. Beih. Ergeb. Limnol.* 10:1–171.

Harris, G. P. 1986. Phytoplankton ecology. Chapman and Hall, London.

Heinemann, H. G., R. F. Holt, and D. L. Rausch. 1973. Sediment and nutrient research on selected corn belt reservoirs. Pages 381–386 in W. C. Ackermann, G. F. White, and E. B. Worthington, eds. Man-made lakes: Their problems and environmental effects. Geophys. Monogr. 17, Am. Geophys. Union, Washington, DC.

Henderson, H. F., R. A. Ryder, and W. Kudhongania. 1973. Assessing fishery potential of lakes and reservoirs. *J. Fish Res. Board Can.* 30:2000–2009.

Higgins, J. M., W. L. Poppe, and M. L. Iwanski. 1981. Eutrophication analysis of TVA reservoirs. Pages 404–412 in H. G. Stefan, ed. Proceedings of the symposium on surface water impoundments. Am. Soc. Civil Engr., New York, NY.

Hoffman, D. A. and A. R. Jonez. 1973. Lake Mead, a case history. Pages 220–223 in W. C. Ackermann, G. F. White, and E. B. Worthington, eds. Manmade lakes: Their problems and environmental effects. Geophys. Monogr. 17, Am. Geophys. Union, Washington, DC.

Hoogenhout, H. and J. Amesz. 1965. Growth rates of photosynthetic microorganisms in laboratory cultures. *Arch. Mikrobiol.* 50:10–25.

Hrbacek, J., M. Dvorakova, V. Korinek, and L. Prochazkova. 1961. Demonstration of the effect of the fish stock on the species composition of zooplankton and the intensity of metabolism of the whole plankton assemblage. *Verh. Internat. Verein. Limnol.* 18:162–170.

Hrbacek, J. and M. Straskraba, eds. 1966. Hydrobiol. Studies 1. Czechoslovak Acad. Sci. 408 pp.

Hrbacek, J. and M. Straskraba, eds. 1973a. Hydrobiol. Studies 2. Czechoslovak Acad. Sci. 348 pp.

Hrbacek, J. and M. Straskraba, eds. 1973b. Hydrobiol. Studies 3. Czechoslovak Acad. Sci. 310 pp.

Hutchinson, G. E. 1957. A treatise on limnology, Vol. 1; Geography, Physics, and Chemistry. John Wiley and Sons, Inc., New York, NY. 1015 pp.

Hutchinson, G. E. 1961. The paradox of the plankton. *Am. Nat.* 95:137–145.

Hynes, H. B. N. 1970. The ecology of running waters. University of Toronto Press, Toronto, Canada. 555 pp.

Hynes, H. B. N. 1975. The stream and its valley. *Verh. Int. Verein. Limnol.* 19:1–15.

Isom, B. G. 1971. Effects of storage and mainstream reservoirs on benthic macroinvertebrates in the Tennessee Valley. Pages 179–192 in G. E. Hall, ed. Reservoir fisheries and limnology. Spec. Publ. 8, Am. Fish. Soc., Washington, DC.

Jassby, A. D. and C. R. Goldman. 1974. Loss rates from a lake phytoplankton community. *Limnol. Oceanogr.* 19:618–627.

Javornicky, P. and J. Komarkova. 1973. The changes in several parameters of plankton primary productivity in Slapy Reservoir, 1980–1967. Their mutual correlations and correlations with the main ecological factors. Pages 155–211 in J. Hrbacek and M. Straskraba, eds. Hydrobiol. Studies 2, Czechoslovak Acad. Sci.

Jenkins, R. M. 1967. The influence of some environmental factors on standing crop and harvest of fishes in U. S. reservoirs. Pages 298–321 in Reservoir fishery resources. Spec. Publ., Am. Fish Soc., Washington, DC.

Jenkins, R. M. 1977. Prediction of fish biomass, harvest, and prey-predator relations in reservoirs. Pages 282–293 in W. Van Winkle, ed. Assessing the effects of power plant-induced mortality on fish populations. Pergamon Press, New York, NY.

Jenkins, R. M. 1982. The morphoedaphic index and reservoir fish production. *Trans. Am. Fish. Soc.* 111:133–140.

Jewson, D. H. and R. B. Wood. 1975. Some effects of integral photosynthesis of artifical circulation of phytoplankton through light gradients. *Verh. Int. Verein. Limnol.* 19:1037–1044.

Jonasson, P. M. 1977. Lake Esrom research, 1867–1977. *Folia Limnol. Scand.* 17:67–89.

Jonasson, P. M. and H. Adalsteinsson. 1979. Phytoplankton production in shallow eutrophic Lake Myvatn, Iceland. *Oikos* 32:113–138.

Jones, J. R. and R. W. Bachmann. 1978. Phosphorus removal by sedimentation in some Iowa reservoirs. *Verh. Int. Verein. Limnol.* 20:1576–1580.

Kalff, J. and R. Knoechel. 1978. Phytoplankton and their dynamics in oligotrophic and eutrophic lakes. *Ann. Rev. Ecol. Syst.* 9:475–495.

Kennedy, R. H., K. W. Thornton, and J. H. Carroll. 1981. Suspended-sediment gradients in Lake Red Rock. Pages 1318–1328 in H. G. Stefan, ed. Proceedings of the symposium on surface water impoundments. Am. Soc. Civil Engr., New York, NY.

Kerr, S. R. 1974. Theory of size distribution in ecological communities. *J. Fish. Res. Board Can.* 31:1859–1862.

Kerr, S. R. 1982. The role of external analysis in fishery science. *Tran. Am. Fish. Soc.* 111:165–170.

Kerr, S. R. and N. V. Martin. 1970. Trophic-dynamics of lake trout population systems. Pages 365–376 in J. H. Steele, ed. Marine food chains. Oliver and Boyd, Edinburgh, Scotland. 552 pp.

Kiefer, D. A., O. Holm-Hansen, C. R. Goldman, R. Richards, and T. Berman.

1972. Phytoplankton in Lake Tahoe: Deep-living populations. *Limnol. Oceanogr.* 17:418–422.

Kiefer, D. A., R. J. Olson, and O. Holm-Hansen. 1976. Another look at the nitrite and chlorophyll maxima in the central North pacific. *Deep-Sea Res.* 23:1199–1208.

Kimmel, B. L. 1981. Land-water interactions: Effects of introduced nutrients and soil particles on reservoir productivity. Tech. Compl. Rept., Proj. No. A-088-OKLA, Office of Water Research and Technology, U.S. Department of Interior. 95 pp.

Kimmel, B. L. 1983. Size distribution of planktonic autotrophy and microheterotrophy: Implications for organic carbon flow in reservoir foodwebs. *Arch. Hydrobiol.* 97:303–319.

Kimmel, B. L. and A. W. Groeger. 1984. Factors controlling phytoplankton production in lakes and reservoirs: A perspective. Pages 277–281 in Lake and reservoir management. EPA 440/5/84-001. U. S. Environmental Protection Agency, Washington, DC.

Kimmel, B. L. and A. W. Groeger. 1986. Limnological and ecological changes associated with reservoir aging. Pages 103–109 in G. E. Hall and M. J. Van Den Avyle, eds. Reservoir fisheries management: Strategies for the 80's. Reservoir Committee, Am. Fish Soc., Bethesda, MD.

Kimmel, B. L. and O. T. Lind. 1972. Factors affecting phytoplankton production in a eutrophic reservoir. *Arch. Hydrobiol.* 71:124–141.

Kimmel, B. L., D. M. Soballe, S. M. Adams, A. V. Palumbo, C. J. Ford, and M. S. Bevelhimer. 1988. Inter-reservoir interactions: Effects of a new impoundment on organic matter production and processing in a mutlipe-impoundment series. *Verh. Int. Verein Limnol.* 23:985–994.

Kimmel, B. L. and M. M. White. 1979. DCMU-enhanced chlorophyll fluorescence as an indicator of the physiological status of reservoir phytoplankton: An initial evaluation. Pages 246–262 in M. W. Lorenzen, ed. Phytoplankton-environmental interactions in reservoirs. U.S. Army Waterways Experiment Station, Vicksburg, MS.

Kimmel, B. L. and A. W. Groeger. 1987. Size distribution of planktonic autotrophy and microheterotrophy in DeGray Reservoir, Arkansas. Pages 297–326 in R. H. Kennedy and J. Nix, eds. Proceedings of the DeGray Lake Symposium. Tech. Report E-87-4, U.S. Army Engineer Waterways Experiment Station, Vicksburg, MS.

Kloet, W. A. de. 1982. The primary production of phytoplankton in Lake Vechten. *Hydrobiologia* 95:37–57.

Knowlton, M. F. and J. R. Jones. 1989. Comparison of surface and depth-integrated composite samples for estimating algal biomass and phosphorus values and notes on the vertical distribution of algae and photosynthetic bacteria in midwestern lakes. *Arch. Hydrobiol./Suppl.* 83:175–196.

Knutson, K. M. 1970. Planktonic ecology of Lake Ashtabula Reservoir, Valley City, North Dakota. Ph.D. Dissertation, North Dakota State Univ., Fargo, N.D. 99 pp.

Lee, G. F., W. Rast, and R. A. Jones. 1972. Eutrophication of water bodies: Insights for an age-old problem. *Env. Sci. Tech.* 12:900–908.

Lelek, A. 1973. Sequence of changes in fish populations of the new tropical man-made lake, Kainji, Nigeria, West Africa. *Arch. Hydrobiol.* 71:381–420.

Liang, Y., J. M. Melack, and J. Wang. 1981. Primary production and fish yields in Chinese ponds and lakes. *Trans. Am. Fish Soc.* 100:346–350.

Likens, G. E. 1972. Nutrients and eutrophication: The limiting nutrient controversy. Spec. Sympos. 1, Am. Soc. Limnol. Oceanogr. 328 pp.

Lind, O. T. 1971. The organic water budget of a Central Texas reservoir. Pages 193–202 in G. E. Hall, ed. Reservoirs fisheries and limnology. Am. Fish Soc.

Lind, O. T. 1979. Reservoir eutrophication: Factors governing primary production. Tech. Compl. Rept., Proj. B-210-TEX, Office of Water Research and Technology, U. S. Department of Interior.

Little, E. C. S. 1966. The invasion of man-made lakes by plants. Pages 75–86 in R. H. Lowe-McConnell, ed. Man-made Lakes. Academic Press, London, England.

Lorenzen, C. J. 1967. Vertical distribution of chlorophyll and phaeopigments: Baja California. *Deep-Sea Res.* 14:735–745.

Lund, J. W. G. 1965. The ecology of the freshwater phytoplankton. *Biol. Rev.* 40:231–293.

Malone, T. C. 1980. Algal size. Pages 433–463 in I. Morris, ed. The physiological ecology of phytoplankton. University of California Press, Berkeley, CA.

Mann, K. H., R. H. Britton, A. Kowalczewski, T. J. Lack, and C. P. Matthews. 1972. Productivity and energy flow at all trophic levels in the River Thames, England. Pages 579–596 in Z. Kajak and A. Hillbricht-Ilkowsa, eds. Productivity of freshwaters. PWN Polish Sci. Publ., Warsaw, Poland.

Margalef, R. 1960. Ideas for a synthetic approach to the ecology of running waters. *Int. Rev. Ges. Hydrobiol.* 45:133–153.

Margalef, R. 1975. Typology of reservoirs. *Verh. Int. Verein. Limnol.* 19:1841–1848.

Marra, J. 1978. Phytoplankton photosynthetic response to vertical movement in a mixed layer. *Mar. Biol.* 46:203–208.

Marra, J. 1978a. Effect of short-term variations in light intensity on photosynthesis of a marine phytoplankter: A laboratory simulation study. *Mar. Biol. (Berl.)* 46:191–202.

Marra, J. 1978b. Phytoplankton photosynthetic response to vertical movement in a mixed layer. *Mar. Biol. (Berl.)* 46:203–8.

Martin, D. B. and R. D. Arneson. 1978. Comparative limnology of a deep-discharge reservoir and a surface-discharge lake on the Madison River, Montana. *Freshwater Biol.* 8:33–42.

Martin, D. B. and J. F. Novotny. 1975. Nutrient limitation of summer phytoplankton growth in two Missouri River reservoirs. *Ecology* 56:199–205.

Marzolf, G. R. 1981. Some aspects of zooplankton existence in surface water impoundments. Pages 1392–1399 in H. G. Stefan, ed. Proceedings of the symposium on surface water impoundments. Am. Soc. Civil Engr., New York, NY.

Marzolf, G. R. and J. A. Osborne. 1971. Primary production in a Great Plains reservoir. *Verh. Int. Ver. Limnol.* 18:126–133.

Marzolf, G. R. and J. A. Arruda. 1980. Roles of materials exported by rivers into reservoirs in the nutrition of cladoceran zooplankton. Pages 53–55 in Restoration of lake and inland waters. U.S. Environmental Protection Agency 440/5-81-010.

Marzolf, G. R. 1984. Reservoirs in the Great Plains of North America. Pages 291–302 in F. B. Taub, ed. Ecosystems of the World 23, Lakes and Reservoirs. Elsevier, New York, NY.

McConnell, W. J. 1963. Primary productivity and fish harvest in a small desert impoudment. *Trans. Am. Fish. Soc.* 92:1–12.

McCullough, J. D. 1978. A study of phytoplankton primary productivity and nutrient concentrations in Livingston Reservoir, Texas. *Tex. J. Sci.* 30:377–387.

McKinley, K. R. and R. G. Wetzel. 1979. Photolithotrophy, photoheterotrophy and chemoheterotrophy: Patterns of resource utilization on an annual and a diurnal basis within a pelagic microbial community. *Microb. Ecol.* 5:1–15.

Megard, R. O. 1981. Effects of planktonic algae on water quality in impoundments of the Mississippe river in Minnesota. Pages 1575–1584 in H. G. Stefan, ed. Proceedings of the symposium on surface water impoundments. Am. Soc. Civil Engr., New York, NY.

Melack, J. M. 1976. Primary production and fish yields in tropical lakes. *Trans. Am. Fish Soc.* 105:575–580.

Meyer, J. L. and G. E. Likens. 1979. Transport and transformation of phosphorus in a forest stream ecosystem. *Ecol.* 60:1255–1269.

Minshall, G. W., R.C. Peterson, K. W. Cummins, T. L. Bott, J. R. Sedell, C. E. Cushing, and R. L. Vannote. 1983. Interbiome comparison of stream ecosystem dynamics. *Ecol. Monogr.* 53:1–25.

Mitchell, S. F. and A. N. Galland. 1981. Phytoplankton photosynthesis, eutrophication and vertical migration of dinoflagellates in a New Zealand reservoir. *Verh. Int. Ver. Limno.* 21:1017–1020.

Naiman, R. J. and J. R. Sedell. 1981. Stream ecosystem research in a watershed perspective. *Verh. Int. Verein. Limnol.* 21:804–811.

National Academy of Sciences. 1969. Eutrophication: Causes, consequences, correctives. Nat. Acad. Sci. Nat. Res. Council, Publ. 1700, Washington, DC. 661 pp.

Neel, J. K. 1963. Impact of reservoirs. Pages 575–593 in D. G. Frey, ed. Limnology in North America. University of Wisconsin Press, Madison, WI.

O'Brien, W. J. 1972. Limiting factors in phytoplankton algae: Their meaning and measurement. *Sci.* 178:616–617.

O'Brien, W. J. 1974. The dynamica of nutrient limitation of phytoplankton algae: A model reconsidered. *Ecol.* 55:135–141.

O'Brien, W. J. 1975. Factors limiting primary productivity in turbid Kansas resrvoirs. Tech. Compl. Rept., Proj. A-0520-KAN, Office of Water Research and Tech., U.S. Department of Interior. 34 pp.

Oglesby, R. T. 1982. The MEI symposium — Overview and observations. *Trans. Am. Fish. Soc.* 111:171–175.

Oglesby, R. T. 1977. Relationships of fish yield to lake phytoplankton standing crop, production, and morphoedaphic factors. *J. Fish. Res. Board Can.* 34: 2271–2279.

O'Neill, R.V., D. L. DeAngelis, J. B. Waide, and T. F. H. Allen. 1986. A hierarchical concept of ecosystems. Princeton Univ. Press, Princeton, J. 253 pp.

Organisation for Economic Co-operation and Development. 1982. Eutrophication of waters — monitoring, assessment, and control. Organisation for Economic Co-operation and Development, Paris, France.

Ostrofsky, M. L. 1978. Trophic changes in reservoirs: An hypothesis using phosphorus budget models. *Int. Rev. Ges. Hydrobiol.* 63:481–499.

Ostrofsky, M. L. and H. C. Duthie. 1978. An approach to modelling productivity in reservoirs. *Verh. Int. Ver. Limnol.* 20:1562–1567.

Ostrofsky, M. L. and H. C. Duthie. 1980. Trophic upsurge and the relationship between phytoplankton biomass and productivity in Smallwood Reservoir, Canada. *Can. J. Bot.* 58:1174–1180.

Parsons, T. R. and J. D. H. Strickland. 1962. On the production of particulate organic carbon by heterotrophic processes in sea water. *Deep-Sea Res.* 8:211–222.

Paulson, L. J., J. R. Baker, and J. E. Deacon. 1979. Potential use of hydroelectric facilities for manipulating the fertility of Lake Mead. Pages 296–300 in

Proceedings of the mitigation symposium, U.S. Dept. Agr. Tech. Rept. No. RM-65.

Paulson, L. J., J. R. Baker, and J. E. Deacon. 1980. The limnological status of Lake Mead and Lake Mohave under present and future powerplant operations of Hoover Dam. Lake Mead Limnol. Res. Cen. Tech. Rept. 1, Univ. Nevada, Las Vegas, NV. 229 pp.

Paulson, L. J. 1981. Nutrient management with hydroelectric dams on the Colorado River system. Lake Mead Limnol. Res. Cen. Tech. Rept. 8, Univ. Nevada, Las Vegas, NV. 39 pp.

Paulson, L. J. and J. R. Baker. 1981. Nutrient interactions among reservoirs on the Colorado River. Pages 1647–1658 in H. G. Stefan, ed. Proceedings of the symposium on surfacewater impoundments. Am. Soc. Civil Engr., New York, NY.

Peterka, J. J. and L. A. Reid. 1966. Primary production and chemical and physical characteristics of lake Ashtabula Reservoir, North Dakota. *Proc., No. Dakota Acad. Sci.* 22:138–158.

Peters, R. H. 1979. Concentrations and kinetics of phosphorus fractions along the trophic gradient of Lake Memphremagog. *J. Fish. Res. Board Can.* 36:970–979.

Peterson, B. J. 1980. Aquatic primary productivity and the ^{14}C-CO_2 method: A history of the productivity problem. *Ann. Rev. Ecol. Syst.* 11:359–385.

Petts, G. E. 1984. Impounded rivers, perspectives for ecological management. John Wiley and Sons, New York, NY.

Placke, J. F. 1983. Trophic status evaluation of TVA reservoirs. Tech. Rept., Tennessee Valley Authority. 136pp.

Placke, J. F. and W. L. Poppe. 1980. Eutrophication analysis of Nickajack and Chickamauga reservoirs. Tech. Rept., Tennessee Valley Authority, Chattanooga, TN. 105 pp.

Platt, T., C. L. Gallegos, and W. B. Harrison. 1980. Photoinhibition of photosynthesis in natural assemblages of marine phytoplankton. *J. Mar. Res.* 38:687–701.

Platt, T. and W. K. W. Li, eds. 1986. Photosynthetic picoplankton. *Can. Bull. Fish. Aquat. Sci.* 214:583 pp.

Ploskey, G. R. 1981. Factors affecting fish production and fishing quality in new reservoirs, with guidance on timber clearing, basin preparation, and filling. U.S. Army Engineer Waterways Experiment Station Tech. Report, Vicksburg, MS. 68 pp.

Poddubny, A. G. 1976. Ecological topography of fish populations in reservoirs. Acad. Sci. USSR, Inst. Biol. Inland Waters. (Translated from Russian and published for the U.S. Bureau of Sport Fisheries and Wildlife, and the National Science Foundation, Amerind Publ. Co., New Delhi.) 414 pp.

Pomeroy, L. R. 1974. The ocean's food web, a changing paradigm. *Bioscience* 24:499–504.

Poppe, W. L., D. J. Bruggink, and J. F. Placke. 1980. Eutrophication analysis of Cherokee Reservoir. Tech. Rep. WR-50-25-80.01. Tennessee Valley Authority, Chattanooga, TN.

Prentki, R. T. and L. J. Paulson. 1983. Historical patterns of phytoplankton productivity in Lake Mead. Pages 105–123 in Aquatic resources management of the Colorado River ecosystem. Ann Arbor Science, Ann Arbor, MI.

Prentki, R. T., L. J. Paulson, and J. R. Baker. 1981. Chemical and biological structure of Lake Mead sediments. Lake Mead Limnological. Res. Cen. Tech. Rept. 6, Univ. Nevada, Las Vegas, NV. 89 pp.

Pridmore, R. D. and G. B. McBride. 1984. Prediction of chlorophyll *a* in impoundments of short hydraulic retention time. *J. Env. Man.* 19:343–350.

Priscu, J. C., J. Verduin, and J. E. Deacon. 1982. Primary productivity and nutrient balance in a lower Colorado River reservoir. *Arch. Hydrobiol.* 94:1–23.

Pyrina, I. L. 1966. Pervichnaya produktsiya fitoplanktona v Inan'-kovskom Rybinskom i Kuibyshevskom vodkhranilischchakh v zavisimosti ot nekotorykh faktorov (Primary Production of phytoplankton in the Ivan'skovskii, Rybinsk, and Kuibyshev reservoirs, in relation to certain factors). Produtsirovanie i krugovorot organisch. veshchestva vo unutr. vodoemakh. Tr. Inst. biol. unutr. vod Akad. Nauk SSSR, No. 13 (16).

Reckhow, K. H. 1979a. Quantitative techniques for the assessment of lake quality. U. S. Environmental Protection Agency Report No. EPA-440/5-79-015. 146 pp.

Reckhow, K. H. 1979b. Empirical lake models for phosphorus: Development, applications, limitations and uncertainty. Pages 193–221 in D. Scavia and A. Robertson, eds. Perspectives on lake ecosystem modeling. Ann Arbor. Sci., Ann Arbor, MI.

Reynolds, C. S. 1984. The ecology of freshwater phytoplankton. Cambridge Univ. Press, Cambridge, England.

Richerson, P., R. Armstrong, and C. R. Goldman. 1970. Contemporaneous disequilibrium, a new hypothesis to explain the "paradox of the plankton." *Proc. Nat. Acad. Sci.* 67:1710–1714.

Richerson, P., M. M. Lopez, and T. Coon. 1978. The deep chlorophyll maximum layer of Lake Tahoe. *Verh. Int. Verein. Limnol.* 20:426–433.

Ridley, J. E. and J. A. Steel. 1975. Ecological aspects of river impoundments.

Pages 565–587 in B. A. Whitton, ed. River ecology. Univ. Calif. Press, Berkeley, CA.

Romanenko, V. I. 1978. Balance of organic matter in the ecosystem of the Rybinskiy Reservoir. Pages 121–131 in Proc., first and second USA-USSR symposia on the effects of pollutants upon aquatic ecosystems, Vol. 1. EPA-600/3-78-076. U. S. Environmental Protection Agency, Washington, DC.

Ryder, R. A. 1965. A method for estimating the potential fish production of north-temperate lakes. *Trans. Am. Fish. Soc.* 94:214–218.

Ryder, R. A. 1978. Ecological heterogeneity between north-temperate reservoirs and glacial lake systems due to differing succession rates and cultural uses. *Verh. Int. Verein. Limnol.* 20:1568–1574.

Ryder, R. A. 1982. The morphoedaphic index - use, abuse, and fundamental concepts. *Trans. Am. Fish. Soc.* 111:154–164.

Ryder, R. A., S. R. Kerr, K. H. Loftus, and H. A. Regier. 1974. The morphoedaphic index, a fish yield estimator - review and evaluation. *J. Fish. Res. Board Can.* 38:663–688.

Salmanov, M. A. and Y. I. Sorokin. 1962. Pervichnaya produktsya Kuibyshevskogo vodokhranilishcha (Primary production from the Kuibyshev reservoir). *Izv. Akad. Nauk SSSR, Ser. Biol.* (4).

Santiago, A. E. 1978. A preliminary study on the primary production of the lower basin, Lake Monroe, Indiana USA. Masters Thesis, Indiana University, South Bend, IN. 117 pp.

Schindler, D. W. 1978. Factors regulating phytoplankton production and standing crop in the world's freshwaters. *Limnol. Oceangr.* 23:478–496.

Schindler, D. W. and S. K. Holmgren. 1971. Primary production and phytoplankton in the Experimental Lakes Area, northwestern Ontario, and other low-carbonate and a liquid scintillation method for determining activity in photosynthesis. *J. Fish. Res. Board Can.* 28:189–201.

Shapiro, J. 1980. The importance of trophic-level interactions to the abundance and species composition of algae in lakes. Pages 105–115 in J. Barica and L. Mur, eds. Hypertrophic systems. Dr. W. Junk, The Hague, The Netherlands.

Sheldon, R. W. and T. R. Parsons. 1967. A continuous size spectrum for particulate matter in the sea. *J. Fish. Res. Board Can.* 24:909–915.

Sheldon, R. W., A. Prakash, and W. H. Sutcliffe, Jr. 1972. The size distribution of particles in the ocean. *Limnol. Oceanogr.* 17:327–340.

Sheldon, R. W., W. H. Sutcliffe, Jr., and M. A. Paranjape. 1977. Structure of pelagic food chain and relationship between plankton and fish production. *J. Fish. Res. Board Can.* 34:2344–2353.

Shiel, R. J. and K. F. Walker. 1984. Zooplankton of regulated and unregulated rivers: The Murray-Darling river system, Australia. Pages 263–270 in A. Lillehammer and S. J. Saltveit, eds. Regulated Rivers. Univ. Oslo.

Sieburth, J. McN., V. Smetacek, and J. Lenz. 1978. Pelagic ecosystem structure: Heterotrophic compartments of the plankton and their relationship to plankton size fractions. *Limnol. Oceanogr.* 23:1256–1263.

Silvey, J. K. G. and J. A. Stanford. 1978. A historical overview of reservoir limnology in the southwestern United States. Pages 1–18 in J. Cairns, E. F. Benfield, and J. R. Webster, eds. Current perspectives of river reservoir ecosystems. North American Benthological Society Publ. 1. Blacksburg, VA. 85 pp.

Smith, R. C., K. S. Baker, O. Holm-Hansen, and R. Olson. 1980. Photoinhibitation of phosynthesis in natural waters. *Photochem. Photobiol.* 31:585–92.

Soballe, D. M., B. L. Kimmel, R. H. Kennedy, and R. M. Gaugaush. In press. Reservoirs. In W. H. Martin, ed. Biotic communities of the Southeastern United States. Ecological Society of America.

Soballe, D. M. 1981. The fate of river phytoplankton in Red Rock Reservoir. Ph.D. Dissertation (Diss. Abstr. No. DA820911), Iowa State Univ., Ames, IA. 92 pp.

Soballe, D. M. and R. W. Bachmann. 1984. Removal of Des Moines River phytoplankton by reservoir transit. *Can. J. Fish. Aq. Sci.* 41:1803–1813.

Soballe, D. M. and B. L. Kimmel. 1987. A large-scale comparison of factors influencing phytoplankton abundance in rivers, lakes, and impoundments. *Ecology* 68:1943–1954.

Soballe, D. M. and S. T. Threlkeld. 1985. Advection, phytoplankton biomass, and nutrient transformations in a rapidly flushed impoundment. *Archiv. Hydro biol.* 105:187–203.

Soltero, R. A. and J. C. Wright. 1975. Primary production studies on a new reservoir; Bighorn Lake — Yellowtail Dam, Montana, U.S.A. *Freshwater Biol.* 5:407–421.

Soltero, R. A., A. F. Gasperino, and W. G. Graham. 1975. Cultural eutrophication of Lone Lake, Washington. *Verh. Int. Ver. Limnol.* 19:1778–1789.

Sonzogni, W. C., S. C. Chapra, D. E. Armstrong, and T. J. Logan. 1982. Bioavailability of phosphorus inputs to lakes. *J. Env. Qual.* 11:555–563.

Sorokin, Yu. I., E. P. Rozanova, and G. A. Sokolova. 1959. Izuchenie pervichnoi producktsii v Gor'kovskom vodokhranilishche s primeneniem ^{14}C (Study of Primary production in the Gorky reservoir using ^{14}C). *Tr. Vsesoyuzn. gidrobiol. obshch.* 9.

Sreenivasan, A. 1972. Energy transformations through primary productivity and fish production in some tropical freshwater impoundments and ponds. Pages

505–514 in Z. Kajak and A. Hillbricht-Ilkowska, eds. Productivity Problems of Freshwaters. PWN Polish Scientific Publishers, Warsaw, Poland.

Stadlemann, P., J. E. Moore, and E. Pickett. 1974. Primary production in relation to temperature structure, biomass concentration, and light conditions at an inshore and offshore station in L. Ontario. *J. Fish. Res. Board Can.* 31:1215–32.

Stanford, J. A. 1978. Relation between plankton dynamics and riverine turbidity in Flathead Lake, MT (Abstr.). *Verh. Int. Verein. Limnol.* 20:1574.

Steele, J. H. 1964. A study of production in the Gulf of Mexico. *J. Mar. Res.* 22:211–222.

Steeman Nielsen, E. 1975. Marine photosynthesis. With special emphasis on the ecological aspects. Elsevier, Amsterdam, Netherlands. 140 pp.

Stockner, J. G. and N. J. Antia. 1986. Algal picoplankton from marine and freshwater ecosystems: A multidisciplinary perspective. *Can. J. Fish. Aq. Sci.* 43:2472–2503.

Straskraba, M., J. Hrbacek, and P. Javornicky. 1973. Effect of an upstream reservoir on the stratification conditions in Slapy Reservoir. Pages 7–82 in J. Hrbacek and M. Straskraba, eds Hydrobiol. Studies 2. Czech. Acad. Sci.

Straskraba, M. and P. Javornicky. 1973. Limnology of two re-regulation reservoirs in Czechoslovakia. Pages 244–316 in J. Hrbacek and M. Straskraba, eds. Hydrobiol. Studies 2. Czech. Acad. Sci.

Stross, R. G. and J. Stottlemeyer. 1965. Primary production in the Patuxent River. *Chesapeake Science* 6:125–140.

Stroud, R. H. and R. G. Martin. 1973. The influence of reservoir discharge location on the water quality, biology, and sport fisheries of reservoirs and tailwaters. Pages 540–558 in W. C. Ackermann, G. F. White, and E. B. Worthington, eds. Man-Made Lakes: Their Problems and Environmental Effects. Geophysical Monograph 17, American Geophysical Union, Washington, DC.

Stuart, T. J. and J. A. Stanford. 1979. A case of thermal pollution limited primary productivity in a Southwestern U.S.A. reservoir. *Hydrobiologia* 58: 199–211.

Sullivan, J. F. 1978. Primary productivity and phytoplankton biomass in the Big Eau Pleine Reservoir, Wisconsin. *Verh. Int. Ver. Limnol.* 20:1581–1586.

Talling, J. F. 1961. Photosynthesis under natural conditions. *Ann. Rev. Plant. Physiol.* 12:133–154.

Talling, J. F. 1971. The underwater light climate as a controlling factor in the production ecology of freshwater phytoplankton. *Mitt. Int. Verein. Limnol.* 19:214–243.

Talling, J. F. and J. Rzoska. 1967. The development of plankton in relation to hydrological regime in the Blue Nile. *J. Ecol.* 55:637–662.

Taylor, H. P. 1971. Phytoplankton productivity responses to nutrients correlated with certain environmental factors in six TVA reservoirs. Pages 209–217 in G. E. Hall, ed. Reservoir fisheries and limnology, Spec. Publ. 8. American Fisheries Society, Washington, DC.

Thornton, K. W., R. H. Kennedy, J. H. Carroll, W. W. Walker, R. C. Gunkel, and S. Ashby. 1981. Reservoir sedimentation and water quality — an heuristic model. Pages 654–661 in H. G. Stefan, ed. Proceedings of the symposium on surface water impoundments. Am. Soc. Civil Engr., New York, NY.

Thornton, K. W. 1984. Regional comparisons of lakes and reservoirs: Geology, climatology, and morphology. Pages 261–265 in Lake and reservoir management. EPA 440/5/84-001. U.S. Environmental Protection Agency, Washington, DC.

Thornton, K. W., R. H. Kennedy, A. D. Magoun, and G. E. Saul. 1982. Reservoir water quality sampling design. *Water Res. Bull.* 18:471–480.

Tilly, L. J. 1975. Changes in water chemistry and primary productivity of a reactor cooling reservoir (Par Pond). Pages 394–407 in F. G. Howell, J. B. Gentry, and M. H. Smith eds. Mineral Cycling in Southeastern Ecosystems. CONF-740513. National Technical Information Service, Springfield, VA.

Tilzer, M. M., C. R. Goldman, R. C. Richards, and R. C. Wrigley. 1976. Influence of sediment inflow on phytoplankton primary productivity in Lake Tahoe (California–Nevada). *Int. Rev. Ges. Hydrobiol.* 61:169–181.

Turner, R. R., E. A. Laws, and R. C. Harris. 1983. Nutrient retention and transformation in relation to hydraulic flushing rate in a small impoundment. *Freshwater Biol.* 13:113–127.

Uhlmann, D. 1968. Der Einfluss der Verweilzeit des Wassersauf die Massenentwicklung von Planktonalgen. *Fortschr. Wasserchem.* 8:32–47.

Vallentyne, J. R. 1974. The algal bowl — lakes and man. Spec. Publ. 22, Dept. of Environment, Ottawa, Canada. 185 pp.

Van Den Avyle, M. J., R. S. Hayward, R. A. Krause, and A. J. Spells. 1982. Spatial variations in abundance of phytoplankton, zooplankton, and larval fishes in Center Hill Reservoir. *Can. J. Water Resources* 7:189–214.

Van Winkle, W., C. C. Coutant, J. W. Elwood, S. G. Hildebrand, J. S. Mattice, and R. B. McLean. 1981. Comparative reservoir research at Oak Ridge National Laboratory. Pages 1432–1447 in H. G. Stefan, ed. Proceedings of the symposium on surface water impoundments. Am. Soc. Civil Engr., New York, NY.

Vannote, R. L., G. W. Minshall, K. W. Cummins, J. R. Sedell, and C. E.

Cushing. 1980. The river continuum concept. *Can. J. Fish. Aquat. Sci.* 37:130–137.

Vincent, W. F. 1979. Mechanisms of rapid photosynthetic adaptation in natural phytoplankton communities. I. Redistribution of excitation energy between photosystems I and II. *J. Phycol.* 15:429–434.

Vincent, W. F. 1980a. Mechanisms of rapid photosynthetic adaptation in natural phytoplankton communities. II. Changes in photochemical capacity as measured by DCMU-induced chlorophyll fluorescence. *J. Phycol.* 16:568–577.

Vincent, W. F. 1980b. The physiological ecology of a *Scenedesmus* population in the hypolimnion of a hypertrophic pond. II. Heterotrophy. *Br. Physol. J.* 15:35–41.

Vincent, W. F. and C. R. Goldman. 1980. Evidence for algal heterotrophy in Lake Tahoe, California-Nevada. *Limnol. Oceanogr.* 25:89–99.

Viner, A. B. 1970. Hydrobiology of Lake Volta, Ghana. II. Some observations on biological features associated with the morphology and water stratification. *Hydrobiologia* 35:230–248.

Volkmar, R. D. 1972. Primary productivity in relation to chemical parameters in Cheat Lake, West Virginia. *Proc. W. Va. Acad. Sci.* 44:14–22.

Vollenweider, R. A. 1975. Input-output models: With special reference to the phosphate loading concept in limnology. *Schwiez. Zeit. fur Hydrol.* 37:53–84.

Vollenweider, R. A. 1976. Advances in defining critical loading levels for phosphorus in lake eutrophication. *Mem. Ist. Ital. Idrobiol.* 33:53–83.

Vollenweider, R. A. and J. J. Kerekes. 1980. Background and summary results of the OECD cooperative program on eutrophication. Pages 25–36 in Restoration of lakes and inland waters, U.S. Environmental Protection Agency, EPA 220-5-81-010.

Walsh, P. and L. Legendre. 1983. Photosynthesis of natural phytoplankton under high frequency light fluctuations simulating those induced by sea surface waves. *Limnol. Oceanogr.* 28:688–697.

Ward, J. V. 1974. A temperature-stressed stream ecosystem below a hypolimnial release mountain reservoir. *Arch. Hydrobiol.* 74:247–275.

Ward, J. V. 1976. Comparative limnology of differentially regulated sections of a Colorado mountain river. *Arch. Hydrobiol.* 78:319–342.

Ward, J. V. and J.A. Stanford, eds. 1979. The ecology of regulated streams. Plenum Press, New York NY. 398 pp.

Ward, J. V. 1981. Tailwater biota: Ecological response to enviornmental alternations. Pages 1516–1525 in H. G. Stefan, ed. Proc. symposium on surface water impoundments. Am. Soc. of Civ. Engineers. New York, NY.

Ward, J. V. and J. A. Stanford. 1983. The serial discontinuity concept of lotic ecosystems. Pages 29–42 in T. Fontaine and S. M. Bartell, eds. Dynamics of lotic ecosystems. Ann. Arbor Science.

Watson, S. and J. Kalff. 1981. Relationships between nannoplankton and lake trophic status. *Can. J. Fish. Aquat. Sci.* 38:960–967.

Webster, J. R., M. E. Gurtz, J. J. Haines, J. L. Meyer, W. T. Swank, J. B. Waide, and J. B. Wallace. 1983. Stability of stream ecosystems. Pages 355–395 in J. R. Barnes and G. W. Minshall, eds. Stream ecology. Plenum Publishing Corp., New York, NY.

Westlake, D. F. 1980. Primary production. Pages 141–246 in E. D. LeCren and R. H. Lowe-McConnell, eds. The functioning of freshwater ecosystems. Cambridge Univ. Press, London.

Wetzel, R. G. 1983. Limnology. W. B. Saunders, Philadelphia, PA. 743 pp.

Wetzel, R. G. 1975b. Primary production. Pages 230–247 in B. A. Whitton, ed. River ecology. Univ. Calif. Press., Berkeley, CA.

Wetzel, R. G. 1983. Limnology. Saunders, Philadelphia, PA.

White, M. M. 1981. Algal heterotrophy in a well-mixed, eutrophic reservoir. M. S. Thesis, University of Oklahoma, Norman, OK. 37 pp.

Williams, P. J. le B. 1981. Incorporation of microheterotrophic processes into the classical paradigm of the planktonic food web. *Kieler Meeresforsch.* 5:1–28.

Woods, P. F. 1981. Physical limnological factors suppressing phytoplankton productivity in Lake Koocanusa, Montana. Pages 1368–1377 in H. G. Stefan, ed. Proc., symposium on surface water impoundments. American Society of Civil Engineers, New York, NY.

Wright, J. C. 1958. The limnology of Canyon Ferry Reservoir. I. Phytoplankton-zooplankton relationships in the euphotic zone during September and October, 1956. *Limnol. Oceanogr.* 3:150–159.

Wright, J. C. 1959. The limnology of Canyon Ferry Reservoir. II. Phytoplankton standing crop and primary productivity. *Limnol. Oceanogr.* 4:235–245.

Wright, J. C. 1960. The limnology of Canyon Ferry Reservoir. III. Some observations on the density dependence of photosynthesis and its cause. *Limnol. Oceanogr.* 5:356–361.

Wright, J. C. 1967. Effects of impoundments on productivity, water chemistry, and heat budgets of rivers. Pages 188–199 in Reservoir fishery resources. Spec. Publ., Am. Fish. Soc., Washington, DC.

Wright, J. C. 1954. The hydrobiology of Atwood Lake, a flood-control reservoir. *Ecol.* 35:305–316.

Young, W. C., H. H. Hannan, and J. W. Tatum. 1972. The physicochemical limnology of a stretch of the Guadalupe River, Texas, with five mainstream impoundments. *Hydrobiologia* 40:297–319.

CHAPTER 7

Reservoirs as Environments for Zooplankton

G. RICHARD MARZOLF

The term plankton refers to the assemblage of organisms living suspended in the open water of lakes and oceans. The term was more general in 1887 when it was coined by Victor Hensen (Welch 1952), and it is used today to refer to the organisms at the mercy of the currents. The term "zooplankton" refers to the animal components of this assemblage. The zooplankton in freshwater is dominated by the arthropod crustaceans and the aschelminth rotifers.

In regions where natural lakes are rare and rivers are the dominant surface water feature, zooplankton are thought to be rare, nonexistent or, at best, ephemeral elements of the fauna. This opinion is widely held (Hynes 1970), but more than likely it is a simple unstated admission that aquatic ecologists have not spent much time observing and collecting in the free-flowing volume of streams and rivers.

Waters (1961, 1962, 1965, 1966) investigated the drifting fauna in streams and found it dominated by macroinvertebrates. Was that because zooplankters were absent, or simply missed because of their size? It does not seem possible that they would have been missed, but that possibility remains open. Chandler (1937) investigated plankton in rivers below lake outflows and found that lake plankton do not survive very far downstream. Cowell (1967) collected zooplankton passing through the outflow structure of a Missouri River reservoir but did not record their fate in the river downstream. Common sense suggests that in order for a zooplankton population to survive at a given spot in a river or stream, the population must be reproducing at a rate equal to or greater than its displacement downstream by the current. In the absence of evidence to the contrary, common sense forces the conclusion that zooplankters do not inhabit flowing water.

The current state of rivers in the United States, and much of the rest of the world for that matter, is such that we probably never will know the presettlement condition of zooplankton assemblages in rivers. Rivers have been used as waste disposal sinks, and more and more of them have been dammed. There have been recent attempts to control the disposal of wastes into rivers, but the future of dam building is not clear. The fact that rivers have been dammed is the central causal fact that explains the appearance of nonflowing habitat for zooplankton in regions where they did not previously exist. As this book is going to press, Saunders and Lewis (1989) document the occurrence of true lentic plankton in the Orinoco River of Venezuela. The source areas for these populations appear to be extensive floodplains when they are in contact with the river at high flows. Nevertheless, at low flow periods this channel itself contains sufficient refuge area for zooplankters.

Zooplankton existence in rivers remains a bit of a mystery because it is observed on occasion, but the focus is on zooplankton in the reservoirs that are impounded when rivers are dammed; the flow is temporarily halted, and zooplankton are now unquestionably capable of reproducing at a rate high enough to maintain populations. Even so, the retention time of water in reservoirs is variable and is likely to influence the fate of zooplankton from time to time.

The perspective of the various elements of reservoir ecosystems provided here underlines the essential differences between ecological events in reservoirs and those in natural lakes that have been studied by more people over a longer period of time. Reservoirs are sufficiently distinct in enough of their basic characteristics to offer novel ways of investigating the adaptations of aquatic organisms.

SOME DIFFERENCES BETWEEN RESERVOIRS AND NATURAL LAKES THAT MIGHT MATTER TO ZOOPLANKTON

Reservoirs, as impoundments of major rivers (sixth to ninth order), differ from natural lakes, where zooplankters are more often studied and where they are certainly better known (Kerfoot 1980). The differences between reservoirs and natural lakes that are of significance for the zooplankton are

1. They are geologically new bodies of water. Since zooplankton populations are not characteristically a component of river ecosystems, the impoundments constructed in the last three or so decades have established new lentic habitats. Dispersal of zooplankton into these habitats was obviously rapid. Does the dispersal process itself influence more rapid changes in species composition in reservoirs than occur in natural lakes that have been in place longer?
   Prophet (personal communication) has observed changes in species composition in Kansas reservoirs over a decade or so. Changes in species composition during the first year or so after impoundment are well known but not well documented. If there is more than one rate involved, what other mechanisms might operate?
2. The catchment areas of reservoirs are much larger relative to their surface areas than are the relative catchment areas of natural lakes. The ratio of drainage area to lake area is greater than 500 for Kansas reservoirs and about 10 for natural lakes in Michigan (Marzolf 1984).
3. The inflow of water from the catchment basin of reservoirs is dominated by the flow from the river(s) impounded. Inflow to natural lakes more often enters from several low-order streams that commonly drain through marshes or extensive littoral deltas before the water becomes the limnetic habitat of the zooplankton. The "filtering" and/or exchange phenomena that influence water quality as it moves through littoral zones may be significant.
4. The volume of many reservoirs is small relative to the discharge of the river into them, and they are shallow relative to the fetch of the wind. Reservoirs behind high dams in mountainous regions are obvious exceptions.

The combined influence of these differences result in features unique to reservoirs in agricultural regions. These are

1. When the drainage area is devoted to row crop agriculture, the export of suspended silts and clays is a general condition that can be dramatically different from water quality conditions in natural lakes

during pulses associated with storm runoff. Suspended load concentrations greater than 1.0 g L^{-1} and Secchi disc disappearance at depths less than 3 cm are not uncommon. Conditions at this extreme are beyond the experience of many limnologists working on natural lakes and for this reason are not often reported.

2. The dominance of a single inflow establishes high concentrations of suspended materials exported by the rivers at the upstream end of the reservoir. As these materials are deposited, a gradient is established along the longitudinal axis of the reservoir. The gradient depends upon the hydrologic regime, (i.e., season), the interval since the last storm runoff pulse, and the operation of the outflow at the dam. The shallowness of the reservoir relative to fetch often precludes stratification and provides for the development of currents sufficient to maintain the particles in suspension; thus, reservoirs are often turbid for their entire length.
3. The turbidity is often sufficient to limit the photosynthetic activity of planktonic algae (Osborne 1972, Marzolf and Osborne 1971). Under turbid conditions the euphotic zone may be limited to the upper meter or so of the water column and, in the absence of stratification, turbulent mixing routinely mixes algal cells to depths where there is insufficient light for their survival. This can have two effects: one, algal biomass is typically low in turbid reservoirs; and two, the algal species that survive these conditions are often facultative or obligate heterotrophs. These conjectures suggest that photosynthesis may be of reduced significance to ecosystem processes in turbid impoundments and indicate that turbid reservoirs offer rich ground for research into algal metabolism in nature.

The appearance of new habitat for zooplankton is coincident with an essentially different set of limnetic conditions for their existence. The dominating feature of the reservoir environment is turbidity, particularly in the U.S. Great Plains, and the presence of longitudinal gradients. The influences of turbid conditions on zooplankton nutrition, species composition, light stimulation of vertical migration, vulnerability to sight-dependent predators, and competitive interactions are all new to the evolutionary history of a cosmopolitan and ubiquitous assemblage of animals. The longitudinal gradient presents an array of particulate resources that is responsible, in large measure and at most times, for variations in population density, species composition, and reproductive performance of zooplankton in reservoirs.

HOW RESOURCE GRADIENTS ARE ESTABLISHED AND HOW THEY CONTROL ZOOPLANKTON POPULATION DENSITY

The evidence is not abundant, but logic suggests that resources are dominant in the control of zooplankton density and species composition. Other controlling mechanisms are possible. Predation was too strongly invoked by Zaret (1980), and Hall (1982) correctly cautioned us. Zaret (1980) was countering the overplayed causality attributed to competition. He argued by similar logic but with little more data. The result was enlightening because it sharpened the issue. Here the issue is opened for the specific case of zooplankton in river impoundments.

The key perspective is the "adversary" nature of individual organism relationships with various elements of their surroundings. Adversary relationships of individuals with the rigors of the physical environment are not only possible, but commonplace. Resource limitation occurs even in the absence of competitors. Adversaries are called competitors only when some resource becomes limiting and another "user" is present. Other users (competitors) might be members of the same species or different species; they may be closely or distantly related.

The competitive exclusion axiom carries the corollary that intensity of competition is related to phylogenetic proximity. This is sound evolutionary logic, but if the environment into which many species have dispersed recently offers an array of new selective pressures, then the logic does not force the conclusion that the intensity of competition is related to phylogeny. Nevertheless, the key issue is that competition must be "for" some part of the available resource array and that to understand competition as an organizing mechanism requires an understanding of the resources.

Predators are adversaries too, with a more definitive effect when they "win"; the prey is eliminated. Nevertheless, the difference in the nature of the relationship as it operates to organize the community is only one of degree. The degree of importance assumed by predation certainly varies, perhaps along gradients in reservoirs, but a consideration of zooplankton resources is a more logical starting place. The following discussion illustrates how features of reservoirs control particulate resources and how these effect the existence of zooplankton. The effect is a dominant one.

THE NATURE OF THE RESOURCE

Nauwerck (1963) realized that phytoplankton alone may not satisfy the energy requirements of natural populations of zooplankton. He forced attention on detritus and bacteria. These alternative food resources have since been shown to be important contributors to zooplankton nutrition (Saunders 1969, Starkweather et al. 1979). Marzolf and Arruda (1981) defined several alternatives in food resources in reservoir conditions where phytoplankton resources are limited by turbidity. Organic detritus enters the reservoir from the river in dissolved and particulate fractions; the dissolved fraction is usually present in greater concentrations, often by as much as 20 times. This has been reported from natural lakes and streams (Wetzel and Rich 1973), from the oceans (Duursma 1960), and from a few rivers (Weber and Moore 1967). The particulate fraction is directly available to filter feeders; the dissolved fraction is not.

Dissolved organic matter is minimally, if at all, useable as food by freshwater filter-feeding zooplankton (Jorgenson 1966). If these compounds were made available by becoming particulate they could represent an important source of carbon in suspension (Sanders 1958). The demonstration of the potential of such a mechanism is incomplete (Robinson 1957). In seawater dissolved organic matter can form organic aggregates on bubbles (Riley 1963, Sutcliffe et al. 1963) that are filterable by *Artemia* (Baylor and Sutcliffe 1963). Buscemi and Puffer (1975) demonstrated that detrital aggregation occurred on calcium carbonate particles that were subsequently ingested by *Bosmina*. Marzolf (1981) demonstrated that a dissolved amino acid (thymidine labeled with radioactive sulfur) could be adsorbed onto clay particles, ingested by daphnids, and subsequently stripped from the clay and assimilated by the animals.

It is not clear that dissolved organic matter can provide the nutrition to maintain zooplankton metabolism, growth, and reproduction by this mechanism in nature. The nutritional qualities of allochthonous dissolved organics are likely to be variable and, in some cases, inadequate since they are the residuals of organic matter decomposition in the terrestrial environment. The use of dissolved organic substrates by bacteria provides, however, an additional mechanism whereby dissolved organic matter can be rendered available to filter feeders. The details of this mechanism remain to be demonstrated, but it is clear that the presence of clay particles enhances the activity of some bacteria (Jannasch and Pritchard 1972, Paerl and Goldman 1972, Goldman and Kimmel 1978). The association of silt and clay particles, dissolved organic matter, and bacteria offers a useable food resource that may be predominantly allochthonous. If so, these interactions link reservoir processes to riparian vegetation and land use in the catchment basin in an important way for zooplankton existence.

One additional comment about the availability of chlorophyll-bearing particulate resources completes the background for consideration of resource use by zooplankton. Concentrations of chlorophyll in midwestern rivers are substantial (50–100 mg m^{-3}). The concentration is proportional to the wetted stream perimeter upstream (Swanson and Bachman 1977). Presumably, benthic algal production serves as a source of cells that is continually eroded and transported by the river. Taylor (1975) found photosynthetic rates of algae in transport to be similar to those of the benthic algal component in the Kansas River. The result is that chlorophyll concentrations are often highest at the inflow end of midwestern reservoirs. Photosynthetic rates in these turbid impoundments are low because the water is deeper in the reservoirs than in the rivers, thus allowing otherwise active cells to be mixed to unlighted depths (Osborne 1972). A large fraction of the riverine algal contribution to such impoundments does not survive passage through the reservoir (Soballe and Bachmann 1984).

EFFECTS OF SUSPENDED SEDIMENTS ON INGESTION RATES AND SURVIVAL OF ZOOPLANKTON

Arruda (1980) and Arruda et al. (1983) document controlled laboratory experiments using sediments and cladocera from Tuttle Creek Reservoir, a turbid reservoir on the Big Blue River in northeast Kansas. Their experiments were conducted to meet three objectives: 1) to evaluate the extent to which suspended clays interfere with the ingestion and incorporation of algae by cladocerans, 2) to measure the ingestion rates of two sizes of suspended sediment particles at several concentrations, and 3) to measure the growth and survival of daphnids fed suspended sediments amended with dissolved organic matter.

Major results were

1. Increasing the sediment concentration decreases rates of ingestion, clearance, and incorporation of *Chlorella vulgaris* by *Daphnia pulex* and *Daphnia parvula*. At suspended sediment concentrations of 100 mg L^{-1} incorporation of algae was reduced by 85% in both species. This implies that in Tuttle Creek Reservoir suspended sediment concentrations as low as 50 mg L^{-1} may reduce the availability of carbon resources to *Daphnia* to starvation levels (Lampert 1977).
2. Ingestion and clearance rates of fine and coarse (mean diameter = 1.88 and 4.65 μm, respectively) clay particles by *Daphnia parvula, Daphnia pulex,* and *Daphnia similis* depended on the size of the particle and the body size and species of the daphnid. Strong inter- and intraspecific differences in the ability to ingest these particles suggest that the particle size distribution might strongly influence resource availability *if* the organic matter adsorbed to clay surfaces is a significant food resource. Small particles have a larger surface, are available for adsorption, *and* are more likely to remain in suspension. It seems clear that the ability to ingest small particles will be important in maximizing the use of the potential resource.
3. Dissolved organic protein adsorbed onto clay particles was fed to cohorts of *Daphnia pulex* and *Daphnia parvula. Daphnia pulex* lived longer and grew larger than sisters fed clean suspended sediments. They lived longer, but were smaller, than controls fed yeast suspensions. *Daphnia parvula* grew equally large and lived equally long on yeast or amended sediments, and this performance was better than sisters fed clean sediments. These experiments were performed with minimal influence of bacterial growth, although it was not absolutely suppressed in the protein-amended treatment. These results show a dissolved organic matter-clay particle-*Daphnia* linkage. However, food levels were low in these experiments and reproductive rates were not impressive. It remains to be seen if naturally occurring dissolved organic materials can be significant as food resources.

Shuman (unpublished data) has measured ingestion rates of suspended sediment particles by *Daphnia* in grazing chambers (Haney 1971) along the turbidity gradient in Tuttle Creek Reservoir. He found that grazing rates are different among the species of filter-feeding zooplankters present. More significantly, he found that ingestion rates varied along the longitudinal axis of the reservoir in direct proportion to concentrations of suspended sediments varying from 2×10^6 to 1.4×10^6 particles ml^{-1}.

A RESOURCE MODEL OF ZOOPLANKTON DISTRIBUTION IN RESERVOIRS

Integration of information about zooplankton existence in reservoirs yields a pattern that suggests that reservoirs behave, to invoke an analogy from chemical engineering, as continuous flow processors. The processes involved are both physical (adsorption and settling) and biological. Materials exported from the catchment basin into the reservoir undergo change, and the water that leaves the reservoir through the outflow structure at the dam is of distinctly different quality. Natural lakes are more analogous to batch processors.

The horizontal distribution (patchiness) of zooplankton populations in lakes is not well known, although Lewis (1980) has recently dealt with the problem in tropical Lake Lanao. The observed horizontal distribution of zooplankton in Tuttle Creek Reservoir is presented in Panel E of Figure 7.1 (Taylor 1971). This pattern is descriptive of the distribution of cladoceran genera *Daphnia, Diaphnanosoma*, and *Bosmina* and of the copepod genera *Diaptomus* and *Cyclops*. This distribution pattern is general and is most likely driven by resources entering the reservoir from the river as they are metabolized by the microflora and ingested by the zooplankton. The hydrodynamics of impoundments place important limits on these patterns.

The reservoir cross-section curve in Panel A (Figure 7.1) is derived from bathymetric data and describes a narrow and/or shallow river becoming wider and/or deeper as the river is impounded. The shape of the curve depends upon the character of the deposits being laid down at the delta and the age of the reservoir. The current velocity curve follows, since a given volume of discharge must move more rapidly through small cross sections than through large ones. Note that water in portions of the reservoir with a small cross section (riverine conditions) will have a

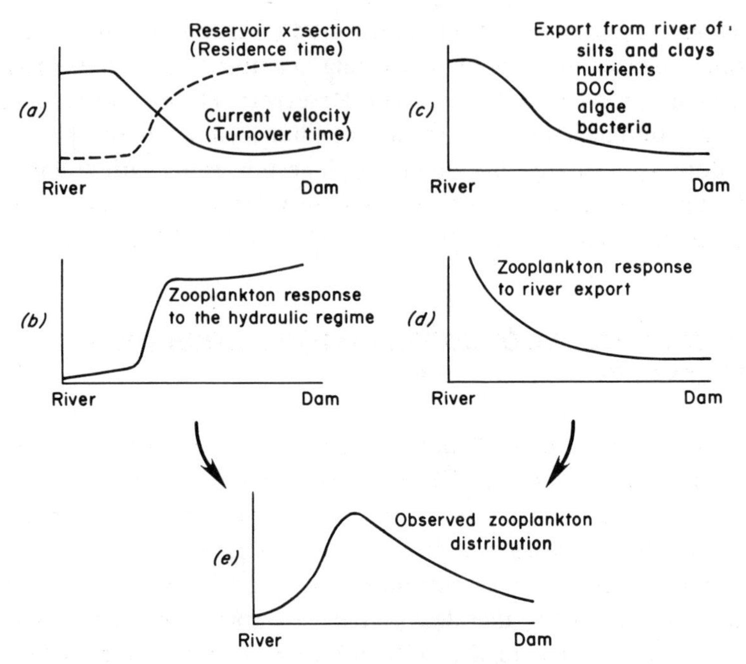

Figure 7.1 An illustration of the several trends that develop along the longitudinal axis of an impoundment, which when taken together result in the observed distribution of zooplankton shown in the bottom panel (E).

short residence time and a rapid turnover time. Water in a portion of the reservoir with a large cross section (impoundment conditions) has a longer residence time and, conversely, the turnover time is reduced.

If current velocity and water turnover time were the only controls, then the pattern in panel B would describe the longitudinal pattern of zooplankton population densities. Under riverine conditions the downstream displacement exceeds the zooplankton reproduction rate. At some point, as water is impounded, the reproductive rate is such that populations can be maintained. This "zooplankton front," if it were examined in detail, would undoubtedly fluctuate upstream and downstream, and species would wax and wane as resources change, a sort of lacustrine aurora borealis.

Panel C in Figure 7.1 illustrates the hypothetical pattern of exports from the river into the reservoir and their fate as they move toward the dam. The decreases downstream result from deposition of particulate materials as current velocity declines and the load-carrying capacity decreases. Nutrients (and contaminants) adsorbed to particles are transported to the sediments by this mechanism. Small particles are maintained in suspension by wind-generated currents or simply by virtue of their colloidal properties. The algae exported from the river into the reservoir suffer losses resulting from limited light availability, settling, or zooplankton grazing (e.g., Soballe and Bachmann 1984). The bacterial distribution shown in Figure 7.1, Panel C, has been observed but is certainly not well understood.

Panel D illustrates the pattern of zooplankton population density as it would appear if the hydraulic effects shown in panel B were not operating. Filter feeding organisms respond to changing quantity or quality of food resources by altering their reproductive performance. The result is that populations increase near the source of food, in this case the river, and the allochthonous sources in its catchment. Vannote et al. (1980) discuss the origin of fine particulate organic matter in flowing water. Their conceptual constructs are consistent with this view of resource export from rivers into reservoirs.

Summation of the curves in panels B and D yields the curve in panel E. The fact that this is the observed distribution is comforting; but then it was knowledge of the distribution that drove the thought process in the first place and prompted the acquisition of the information. Logic demands that this must be the distribution if the proposed model is true, but the fact that the observed distribution "fits" the prediction of the model is not sufficient proof of its truth. There are experiments, yet to be designed and performed, that will provide more or less confidence in this idea.

The lower limits of particle size that can be filtered and ingested by zooplankton are still unknown. The abilities of filter feeders to discriminate among particles is not known. The qualities of dissolved organic matter and how those influence microbial populations in reservoirs is unknown. An improved understanding of the role of these resources in the nutrition of zooplankton and the reciprocal role of zooplankton in the control of these water quality variables is needed. This knowledge is central to understanding how biological processes control the quality of water.

ACKNOWLEDGMENTS

Ed LaRow, Mel Taylor, Joe Schwartz, Don Dufford, Joe Cramer, Mike Novak, John Osborne, Tom Horst, Mitch Taylor, John Shuman, Robin Faulk, Ken Kemp, and Steve Fretwell have all, at one time or another or in one way or another, contributed to the development of these ideas. Joe Arruda has recently been central to the organization of the current statement. Discussions with other authors in this volume and colleagues attending meetings of the Great Plains Limnological Association, especially Roger Bachmann, Dale Toetz, and Troy Dorris, have provided continuing stimulation. The Kansas State University Agricultural Experiment Station has provided facilities and support, and so has the National Science Foundation.

REFERENCES

Arruda, J. A. 1980. Some effects of suspended silts and clays on the feeding of *Daphnia spp.* from Tuttle Creek Reservoir. Ph.D. Thesis, Kansas State University, Manhattan, KS.

Arruda, J. A., G. R. Marzolf, and R. T. Faulk. 1983. The role of suspended sediments in the nutrition of zooplankton in turbid reservoirs. *Ecology* 64: 1225–1235.

Baylor, E. R. and W. H. Sutcliffe, Jr. 1963. Dissolved organic matter in sea water as a source of particulate food. *Limnol. Oceanogr.* 8:369–371.

Buscemi, P. A. and J. H. Puffer. 1975. Chemotrophic attributes of detrital aggregates in a New Mexico alkaline reservoir. *Verh. Verein. Int. Limnol.* 19:358–366.

Chandler, D. C. 1937. The fate of typical lake plankton in streams. *Ecol. Monogr.* 7:445–479.

Cowell, B. C. 1967. The copepoda and cladocera of a Missouri River Reservoir: A comparison of sampling in the reservoir and the discharge. *Limnol. Oceanogr.* 12:125–136.

Duursma, E. K. 1960. Dissolved organic carbon, nitrogen, and phosphorus in the sea. *Neth. Sea Res.* 1:1.

Goldman, C. R. and B. Kimmel. 1978. Biological processes associated with suspended sediment and detritus in lakes and reservoirs. In J. Cairns, E. F. Benfield, and J. R. Webster, eds. Current perspectives on river reservoir ecosystems. North American Benthological Society, Publication No. 1.

Hall, D. J. 1982. Review. *Limnol. Oceanogr.* 27:391–393.

Haney, J. F. 1971. An *in situ* method for measurement of zooplankton grazing rates. *Limnol. Oceanogr.* 13:476–484.

Hynes, H. B. N. 1970. The ecology of running waters. University of Toronto Press, Toronto, Canada. 555 pp.

Jannasch, H. W. and P. H. Pritchard. 1972. The role of inert particulate matter in the activity of aquatic microorganisms. *Mem. Ist. Ital. Idrobiol. Suppl.* 29:289–308.

Jorgenson, C. B. 1966. Biology of suspension feeding. Pergammon Press, New York, NY. 313 pp.

Kerfoot, W. C., ed. 1980. Evolution and ecology of zooplankton communities. University of New England Press, Hanover, NH. 793 pp.

Lampert, W. 1977. Studies on the carbon balance of *Daphnia pulex* as related to environmental conditions. IV. Determination of the "threshold" concentration as a factor controlling the abundance of zooplankton species. *Arch. Hydrobiol. Suppl.* 48:361–368.

Lewis, W. M. 1980. Evidence for stable zooplankton community structure gradients maintained by predation. Pages 625–634 in W. C. Kerfoot, ed. Evolution of ecology and zooplankton communities, University of New England Press, Hanover, NH.

Marzolf, G. R. 1981. Some aspects of zooplankton existence in surface water impoundments. Pages 1342–1388 in H. Stefan, ed. Proceedings of the symposium on surface water impoundments. Amer. Soc. Civil Engr., Minneapolis, MN.

Marzolf, G. R. 1984. Reservoirs in the great plains of North America. In F. B. Taub, ed. Lake and reservoir ecosystems. Elsevier Sci. Publ. Co., Amsterdam.

Marzolf, G. R. and J. A. Arruda. 1980. Roles of materials exported by rivers into reservoirs in the nutrition of cladoceran zooplankton. Pages 53–58 in Restoration of lakes and inland waters. EPA 440/5-81-010, United States Environmental Protection Agency.

Marzolf, G. R. and J. A. Osborne. 1972. Primary production in a Great Plains Reservoir. *Verh. Ver. Int. Limnol.* 18:126–133.

Nauwerck, A. 1963. Die Beziehungen zwischen zooplankton und phytoplankton. *See Erken. Symb. Bot. Uppsala.* 17:1–163.

Osborne, J. A. 1972. The application of a photosynthetic model for turbid reservoirs: A field investigation. Ph.D. Thesis, Kansas State University, Manhattan, KS.

Paerl, H. W. and C. R. Goldman. 1972. Stimulation of heterotrophic and autotrophic activities of a planktonic microbial community by siltation at lake Tahoe, California. *Mem. Ist. Ital. Idrobiol. Suppl.* 29:129–147.

Riley, J. A. 1963. Organic aggregates in seawater and the dynamics of their formation and utilization. *Limnology and Oceanogr.* 8:372–381.

Robinson, M. 1957. The effects of suspended materials on the reproductive rate of *Daphnia magna. Publ. Inst. Mar. Sci.* 4:265–277.

Sanders, H. L. 1958. Benthics studies in Buzzards Bay. I. Animal-sediment relationships. *Limnol. Oceanogr.* 3:245–258.

Saunders, G. W. 1969. Some aspects of feeding in zooplankton. Pages 556–573 in Eutrophication: Causes, consequences, and correctives. National Academy of Sciences, Washington, DC.

Saunders, J. F., III and W. M. Lewis. 1989. Zooplankton abundance in the lower Orinoco River, Venezuela. *Limnology and Oceanography* 34(2):397–409.

Soballe, D. M. and R. W. Bachmann. 1984. Influence of reservoir transit on riverine algal transport and abundance. *Canadian Journal of Fisheries and Aquatic Sciences* 41:1803–1813.

Starkweather, P. L., J. J. Gilbert, and T. M. Frost. 1979. Bacterial feeding by the rotifer, *Brachionus calyciflorus* clearance and ingestion rates, behavior and population dynamics. *Oecologia* 44:26–30.

Sutcliffe, W. H., Jr., E. R. Baylor, and D. W. Menzel. 1963. Sea surface chemistry and Langmuir circulation. *Deep Sea Research* 10:233–243.

Swanson, C. D. and R. W. Bachmann. 1976. A model of algal exports in Iowa streams. *Ecology* 57:1076–1080.

Taylor, M. K. 1975. Photosynthesis in the Kansas River. Kansas State University, Manhattan, KS.

Taylor M. W. 1971. Zooplankton ecology of a Great Plains Reservoir. M. S. Thesis, Kansas State University, Manhattan, KS.

Waters, T. F. 1961. Standing crop and drift of stream bottom organisms. *Ecology* 42:532–537.

Waters, T. F. 1962. Diurnal periodicity in the drift of stream invertebrates. *Ecology* 43:316–320.

Waters, T. F. 1965. Interpretation of invertebrate drift in streams. *Ecology* 46:327–334.

Waters, T. F. 1966. Production rate, population density, and drift of a stream invertebrate. *Ecology* 47:595–604.

Weber, C. I. and D. R. Moore. 1967. Phytoplankton seston and dissolved organic matter in the Little Miami River at Cincinnati, Ohio. *Limnol. Oceanogr.* 12:311–318.

Welch, P. S. 1952. Limnology. McGraw-Hill, New York, NY. 538 pp.

Wetzel, R. G. and P. H. Rich. 1973. Carbon in freshwater systems. In G. M. Woodwell and E. V. Pecan, eds. Carbon and the biosphere. Proc. Brookhaven Symposium in Biology. 24 pp.

Vannote, R. L., G. W. Minshall, K. W. Cummins, J. R. Sedell, and C. R. Cushing. 1980. The river continuum concept. *Can. J. Fish. Aquat. Sci.* 37:130–137.

Zaret, T. M. 1980. Predation and freshwater communities. Yale University Press, New Haven, CT. 187 pp.

CHAPTER 8

Perspectives on Fish in Reservoir Limnology

W. JOHN O'BRIEN

Reservoirs in North America have been built in greatest profusion in areas where natural lakes are sparse. Without impoundments these regions would have little standing water for recreation and greater danger of flooding as all runoff flows directly into streams and rivers. Because glaciers were the prolific builders of natural lakes, both large and small, in North America, reservoirs tend to have been built in those areas that have not been recently glaciated: the Midwest, Southeast, and Southwest. For example, Oklahoma has 67 reservoirs greater than 500 acres at average pool levels, with a total area of 557,000, while Minnesota has only 12 large reservoirs, with a total area of 62,000 acres (Ploskey and Jenkins, 1980).

Naturally, the geographical distribution of reservoirs also dictates a good bit about their biology. Many reservoirs are shallow, and they tend to be in warm climates. This combination virtually assures a fish composition excluding most cold water fish, such as the salmonids, trouts, and white fishes, while including most warm water fish, such as the centrachids, ictalurids, and many cyprinids. However, fish species composition in reservoirs does not seem to differ strikingly from that occuring in natural lakes located in similar latitudes.

While the species composition of reservoirs and natural lakes may be similar, there are often differences in the relative density distribution of particular species. These differences are partly related to reservoir stocking. Almost every reservoir of any size is initially stocked with fish species thought to be beneficial either for sport or as forage for sport fish.

Commonly stocked sport fish are largemouth bass, white bass, walleye, flathead catfish, channel catfish, crappie, and some exotics, such as striped bass, a marine fish. Other species are stocked to provide forage for these largely piscivorous sport fish. The ideal forage species should be very fecund, have stable population growth characteristics, be trophically efficient, be vulnerable to predation, and hopefully efficient, be vulnerable to predation, and hopefully be innocuous (Ney 1981). Obviously, no one species can be all these things.

No species has been more widely stocked for forage than the bluegill sunfish, although a variety of other sunfish are either stocked as forage or commonly occur in the impounded streams and thus get into the reservoirs. Various minnows are also stocked or are so commonly used as bait fish that fisherman soon stock them inadvertently; these include the golden shiner and fathead minnow. Gizzard shad and threadfin shad are very commonly stocked, although fisheries biologists have been split in their opinions as to the value of gizzard shad as a forage fish. In reservoirs that reach 5°C or freeze over, threadfin shad will not survive and thus must be restocked each year (Griffith 1978). However, some fisheries biologists have exploited this fact and have stocked threadfin shad rather than gizzard shad, which can ultimately grow to a rather large size and thus become unavailable as forage to sport fish. Because threadfin shad cannot reach such size in just one year they always remain edible. Stocking of alewives, blueback and Atlantic herring, and rainbow smelt has been limited to reservoirs in southeastern North America, close to the coast.

As has been emphasized in previous chapters, reservoirs have some chemical and physical characteristics that are somewhat different from natural lakes. Still, fish communities are often quite similar in the two types of water bodies. It must be, then, that despite some differences between lakes and reservoirs, reservoirs provide all the essential components of the niches of many warm-water fishes. What these essential components are is still a matter of considerable discussion and research, but certainly some general characteristics are known. To sustain fish growth and reproduction any water body must provide habitats that meet the chemical and physical requirements of a fish species. When fish species are unable to survive in impounded water bodies it is often for lack of suitable water temperature and/or dissolved oxygen, diversity in habitat, unsuitable spawning sites, lack of sufficient prey for a particular stage in the life cycle, or lack of refuge from predators.

Factors Affecting Spawning Success

Changing water levels and substrate composition. It has been quite clear for years that for any nest-building fish or any fish in which the eggs attach to a particular substrate the nature of the substrate is important in successful spawning. Reservoirs built in old, highly weathered landscapes may have poor bedrock materials to provide stable submerged substrates needed for spawning by such fish as walleyes. The importance of vegetation for crappie spawning has long been known (Hansen 1951). Vegetation may also be necessary and certainly enchances spawning success for other centrachids and yellow perch and bigmouth buffalo (Johnson 1963). However, soil type and wave action, unconsolidated shorelines, and high levels of turbidity often preclude the development of substantial littoral zone vegetation, and thus these reservoirs do not provide good spawning habitat for species dependent upon vegetation.

A common attribute of reservoirs is that the water level fluctuates greatly. This can have both positive and negative effects on spawning, which for most warm-water fish is in the springtime. Because most gravel beds occur in the upper few meters of the littoral zone, where wave elutriation keeps gravel beds free of silt, drawdown below this level may leave only muddy areas that are poor for spawning. Drawdown also minimizes the amount of vegetation available for spawning, especially in turbid reservoirs, where light sufficient for macrophytic growth may penetrate only the top meter or so.

Increase in water level, which is common in the often rainy spring of the midwestern and southeastern United States, may have just the opposite effect. Spawning success has been found to increase either as reservoirs fill and thus flood vegetation or in years of high water (Martin et al. 1981, Walburg 1977, Martin and Campbell 1953, Shirley and Andrews 1977). High water sometimes floods terrestrial vegetation and could well inundate gravel areas created through erosion around the reservoir shore. These events provide the diversity of substrates needed for spawning.

Siltation. Watersheds within our area of major interest are comprised mainly of fine silts and clays; thus, reservoirs are characterized by moderate to heavy turbidity and siltation. Silt deposition on fish eggs increases egg mortality and thus decreases spawning success (Ploskey 1981). It has been shown that eggs of northern pike, when covered by silt at a rate of 1 mm per day, had only 3% survival (Hassler 1970).

Littoral zone vegetation and structure. As mentioned previously, one of the major differences between many natural lakes and most reservoirs is the rather unstructured shoreline and littoral zone of reservoirs and the sparsity of littoral vegetation as compared with natural lakes. Structure is known to be important for a variety of nest-building fish. Largemouth bass rarely nest on sand or silt but rather close to large rocks or ledges (Allan and Romero 1975). Kramer and Smith (1962) found that almost all the largemouth bass nests in Lake George, Minnesota, were situated in macrophytic vegetation, and black crappie almost always deposit their eggs at the base of aquatic vegetation (Ginnelly 1971).

Aquatic vegetation thus often provides the very substrate within which or on which eggs are laid (Benson 1976) and may also protect eggs from wave action and erosion. The role of nearby structure is less clear, but it doubtless makes nest defense from egg predators more effective (Vogele 1975). Some have tried providing artificial cover in reservoirs but have met with only mixed success (Clady and Summerfelt 1979).

Factors Affecting Larval Fish and Fingerling Survival

Food acquisition. Survival of larval fish is dependent primarily upon acquiring food and avoiding predation. The idea that there is a critical period between utilization of the yolk sac and initial feeding on zooplankton has a long history (Fabre-Domerque and Bietrix 1897). Almost all larval fish feed on zooplankton. However, as fish are very small in larval stages, their early feeding may be gape-limited (Zaret 1981). That is, larval fish may be able to feed only on the tiniest of zooplankton. Thus, early growth and survival of fish may depend upon the densities of small cladocerans and rotifers. The state of gape-limited feeding is brief. Hunter (1979) found that marine fishes could feed on the entire spectrum of zooplankton sizes available in the environment within three weeks of hatching, and Hansen and Wahl (1981) found that once yellow perch reached 3.0 cm, they could handle the largest *Daphnia pulex* available (2.2 mm) (Figure 8.1).

Predation. Survival of larval fish is probably influenced more by predation than by feeding. These very small fish are vulnerable to virtually every other predator in lakes and reservoirs. Not only visual-feeding fish

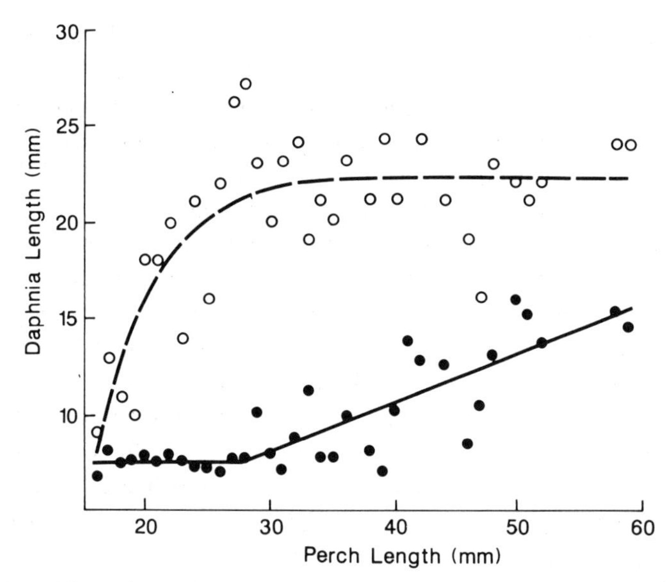

Figure 8.1 The relationship of yellow perch length to median length (solid circles) and maximum length (open circles) *Daphnia pulex* consumed in Oneida Lake, New York. The solid line is a splined regression of median length *D. pulex* to yellow perch length, and the dashed line is an asymptotic curve describing maximum length of *D. pulex* in yellow perch stomachs. (After Hansen and Wahl 1981.)

but other predators such as predaceous copepods may have considerable influence on larval fish densities. *Mesocyclops leucartii* and *Acanthocyclops vernalis* have been found to attack larval *Alosa* species (Nikolsky 1963), and several *Cyclops* species have been observed to attack and consume smelt larvae (Lillelund 1967).

Protective cover, such as aquatic macrophytes, must be especially critical in minimizing fish predation on small fish. Savino and Stein (1982) showed that largemouth bass predation success on bluegill sunfish declined dramatically as simulated, submersed vegetation density was increased to 250 stems/m^2. Any factor(s) that would reduce vegetation cover, such as turbidity or wave action, could also minimize larval fish survivorship. Drawdown of water levels below the vegetation zone would seem to be especially disastrous to larval fish. Aggus and Elliot (1975) found that largemouth bass young-of-the-year survivorship was increased when terrestrial vegetation was flooded in Bull Shoals Lake.

Structural complexity, especially aquatic vegetation, while providing refuge for larval and fingerling fish, may also reduce the ability of piscivorous fish to feed on smaller fish (Cooper and Crowder 1979). Colle and Shireman (1980) found that condition factor greatly increased for largemouth bass, redear, and bluegill sunfish when the macrophyte *Hydrilla* declined in several lakes in Florida. Thus, fisheries biologists face a dilemma in that vegetation which increases forage fish survivorship may ultimately reduce sport fish production. Fisheries biologists have exploited the capability to draw down reservoirs later in the year, in late summer and early fall, to force very vulnerable young-of-the-year forage fish out of vegetation structure and thus make them accessible to predators (Keith 1975). Heman (1965) found much greater growth of medium and large largemouth bass after such a drawdown. However, summer drawdown may decimate the younger year-class of most fish species (Noble 1981).

Factors Affecting Fish Feeding

Ecologists have long recognized that the rate and manner of feeding is of paramount importance to the success of almost all animals. As Fabre (1913) said, "From the least to the greatest in the zoological progression, the stomach sways the world; the data supplied by food are chief among all the documents of life." This is certainly true of fish. Over the years a great deal of research has been devoted to describing fish diet composition and feeding rate. Ivlev (1961) was one of the first fish ecologists to quantify accurately the components of fish feeding ecology. He identified two major factors that need to be addressed: the preference fish may have for various prey, and the availability of prey to the fish at the time it feeds. This last point is often overlooked in field studies of fish feeding because matching up the location of fish in a lake or reservoir and the prey available at those times and places is difficult.

Using Ivlev's formulation, however, any consideration of fish feeding ecology in a water body must first address where the fish are in the lake, since this largely determines the prey species and types available to the fish. The distribution of fish in lakes and reservoirs has been studied extensively with a variety of techniques. However, each of the techniques in common use has limitations that make it less than ideal for studying fish feeding. For example, many types of traps, gill nets, and trawling devices work best, or sometimes only, during the night, when fish cannot see to avoid them. Because many fish are daytime or crepuscular feeders, these sampling techniques do not help determine where fish are when they feed.

Within-reservoir fish distribution. Despite the limitations of presently used sampling methods considerable information on the usual location of reservoir fish is available. Vertically strung gill nets placed in the pelagia and midwater trawls have convincingly shown that the major pelagic fish in reservoirs are gizzard and threadfin shad, white and black crappie, often yellowfin perch, and walleye. Other than small black and white crappie, few sunfish are found in the pelagia of large reservoirs. Bottom trawling often catches a variety of catfishes, along with carp, carpsuckers, and buffaloes. Trap nets, if the stringer line extends very far off shore, may catch some of the pelagic species mentioned above but will include almost all of the littoral sunfishes, and most of the minnows as well.

As is easily appreciated, all of these methods have limitations in studying fish feeding. Trawls and seines catch fish over a large area, and gill nets and trap nets are set out for long periods of time. In neither case can detailed knowledge of the microhabitat distribution of fish be obtained.

Other sampling methods have, of course, been used, and to good purposes, but generally not to demonstrate the daytime or feeding distributions of fish. Electroshocking is a very effective method of catching fish; however, its effectiveness is limited to the upper few meters of a reservoir, and it, too, is generally most effective at night. Sonar sampling has become very sophisticated, but it is usually not possible to determine even the size of fish, let alone species, from sonar traces. Furthermore, sonar simply cannot function in shallow or weedy areas, where many fish are found.

What we do know of the microdistribution of warm-water fish comes from a few scuba studies in clear northern lakes. Werner et al. (1977) visually observed eight warm-water species of fish, including sunfish, yellow perch, and shiners, in the spring and summer. They found less habitat segregation in the spring than in the summer, but in summer they found very consistent distributions. Green sunfish were almost always found very near shore, while black crappie were always found in the open water. Both bluegill and largemouth bass were found throughout the littoral and open water zones but rarely right near shore. A similar distribution most likely exists in many reservoirs that have these same species, but there is little direct evidence for this. Reservoir trawling, netting, and seining commonly find fish in areas not inconsistent with this more precise distribution information.

Reservoir fish feeding. Fish that live in the open waters of lakes and reservoirs must feed primarily on zooplankton, perhaps some on phytoplankton, whereas littoral fish may choose from a more diverse prey array, including terrestrial insects, macrophytic vegetation, benthic invertebrates, and forage fish. Although in anything but a small pond terrestrial insects may be trivial to the overall level of fish production, during the summer they may be important sources of food right along the shore line, particularly for green sunfish.

Zooplankton in the pelagia of lakes and reservoirs probably make up the vast majority of biomass potentially available to fish. Almost all fish species exploit zooplankton as food as larval fish and fingerlings, but only a few remain planktivorous throughout their entire life cycle, probably because zooplankton are such a small, diffuse food. Black and white crappie have been shown to be primarily zooplanktivorous until 12–16 cm total length (*TL*), probably corresponding to the second or third year, at which time they become more littoral in distribution and feed on benthos and other fish. Gizzard shad are also pelagic, especially within this same size range, and feed on both zooplankton and larger phytoplankton. Drenner et al. (1982) showed that gizzard shad feed by pumping or gulping water into the buccal cavity and straining it through the very fine gill rakers (average distance between rakers is 80–100 mm). This appears to be the major foraging mode of gizzard shad after the fingerling stage, and certain pelagic whitefish have been shown to show this same behavior occasionally (Janssen 1976). In more southern reservoirs various artherinid planktivores also feed on zooplankton in the pelagia.

Aquatic macrophytes are little used as a food source by any lake-dwelling organisms and used hardly at all by fish native to North America. Benthic feeding omnivores, such as carps and buffaloes, may utilize energy initially produced by macrophytes, but only after the plants have died in the autumn and the leaves and stems have begun to decompose. However, the grass carp, an exotic fish, is often stocked because of its voracious feeding on macrophytes. It is being widely introduced throughout the southern United States (Guillory and Gasamay 1978), although many of the potential ramifications of this type of feeding have not been worked out.

A major food resource for fish in many natural lakes are the various benthic invertebrates, snails, and aquatic insects that inhabit the lake bottom and, to some extent, the vegetation of the littoral zone. It is on this food resource that most littoral fish feed. All the littoral sunfish utilize these food sources to some extent, some species almost exclusively. Largemouth bass feed primarily on benthos until they reach 6–10 cm in size.

Of most interest to fishermen and fisheries biologists are those fish that feed on other fish. Largemouth bass are a major widespread piscivore, as are walleye, white bass, and large white and black crappie. The largemouth bass and crappie do much of their feeding near the littoral zone. When gizzard shad are present in a reservoir, they may take up 50–80% of the diet composition of largemouth bass, yet shad are thought to be pelagic and bass littoral. It is not known whether the bass venture into the pelagia to feed on the shad, or whether they catch the occasional shad or part of a shad school that comes near to shore. This strong selection for shad is presumed to be related to the ease of handling of these soft-rayed fishes, as opposed to the difficulty of eating spiny fish such as the sunfish. Still, shad can become a problem because they can often develop high densities of large individuals that cannot be fed upon by bass and may interfere with the production of other species (Noble 1981). Surprisingly, as often as bluegill are stocked as forage for largemouth bass, most studies find that bass will eat most other species before feeding on bluegill.

Howick and O'Brien (1983) studied in detail the feeding ecology of largemouth bass. They found that the distance at which bass could locate forage fish prey varied from little more than 40 cm for small bluegill sunfish (3 cm) to 200 cm for large bluegill (8 cm TL). These distances were measured under high light intensity at 3,000 lux, and the distances declined greatly as light intensity fell below 5 lux (Figure 8.2). However, at these low light intensities bass could often locate the larger sunfish at greater distances than the sunfish could locate the approaching bass. This was not so true for small size (Figure 8.2). Fisheries biologists have long recognized that forage fish can grow to such a size that piscivorous fish an no longer handle and ingest them (Lawrence 1957), but Howick and

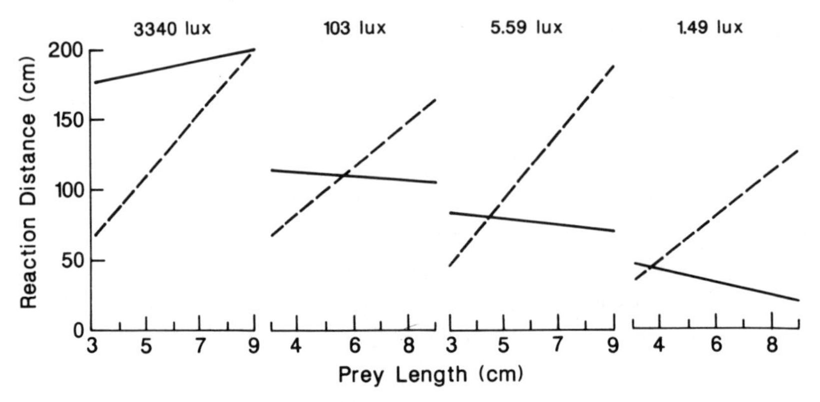

Figure 8.2 The reciprocal visual abilities of bluegill sunfish and largemouth bass in locating the other under different light intensities. The dashed line indicates the reaction distance of a 29-cm largemouth bass locating different-sized bluegill. The solid line indicates the startle distance of different-sized bluegill locating a 29-cm largemouth bass. Bluegill smaller than the intercept of the two lines should be relatively safe from bass predation by being able to locate the bass before the bass can locate that size prey. (After Howick and O'Brien 1983.)

O'Brien (1983) showed that there is also a small size where the forage fish can locate and flee from a piscivore well before the piscivorous fish can locate and attack the forage fish.

The Effects of Turbidity on the Feeding of Reservoir Fish

High levels of suspended silts and clays are one of the most consistent and pervasive factors in which reservoirs differ from natural lakes. As we have seen, these silts and clays have effects at all trophic levels. At worst, suspended silts and clays can clog fish gills and make fish life impossible, but silts and clays have considerable impact long before this effect is realized. As had been pointed out previously (see Chapter 6), silts and clays greatly reduce the euphotic zone within a reservoir. Many natural lakes have euphotic zones several tens of meters deep, yet the euphotic zone of even a moderately turbid reservoir will be only a few meters deep at best. Turbidity thus severely restricts the zone within the lake where visually feeding fish can efficiently find and attack their prey.

Turbidity also reduces fish vision within the euphotic zone. Vinyard and O'Brien (1976) found that turbidities of 30 JTU's, commonly

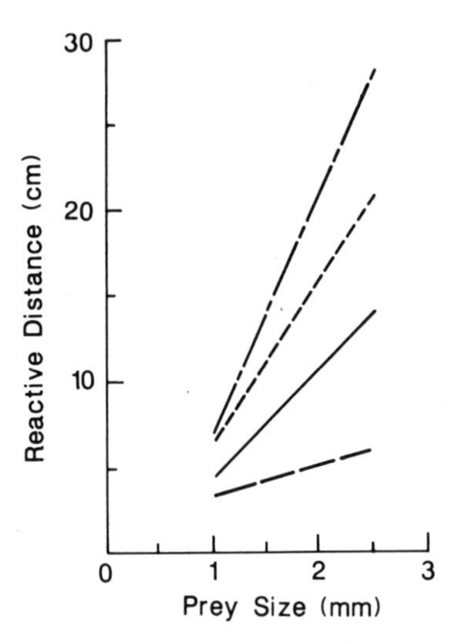

Figure 8.3 The relationship between reactive distance of bluegill sunfish and different-sized *Daphnia pulex* for different turbidities at a constant illuminance (34.9 lx). The long-short dashed line indicates the 6.3-JTU observations. The short dashed line indicates the 10-JTU observations. The solid line indicates the 20-JTU observations. The long dashed line indicates the 30-JTU observations. JTU stands for Jackson turbidity units. (After Vinyard and O'Brien 1976.)

observed in many reservoirs, reduced the reactive distance of bluegill sunfish to a 2-mm *Daphnia pulex* to 5 cm, as compared to a reactive distance for this same species and size prey at low turbidity and modest light intensity of 25 cm (Figure 8.3). This indicates a fivefold decrease in reactive distance between 2–3 and 30 JTU's. However, the actual impact on bluegill foraging is even greater because the reactive distance represents only the radius of a search volume that is hemispheric. Therefore, a fivefold decrease in reactive distance means that the search volume is reduced more than 100 times. If feeding rates are dependent upon search volume, such dramatic decreases in prey visibility may almost preclude visual feeding by planktivorous fish at these and higher turbidity levels.

Janssen (1978) and Drenner et al. (1982) have pointed out that there are several species of fish that feed on planktonic organisms without using visual location; they either strain water through their gills while swimming or, more commonly, pump water through the gills to strain out planktonic organisms. This latter mode of feeding is utilized by the clupeids, gizzard and threadfin shad, and others. This mode of feeding is presumably much less affected by high turbidity levels. These fish may therefore be expected to increase with turbidity levels. There is no doubt that in many reservoirs the dominant pelagic fish is gizzard shad. Drenner and McComas (1980) have shown that zooplankton with excellent current evasion abilities such as those of diaptomid copepods and *Diaphanosoma* may be favored as these pump filter-feeding fish become abundant.

Fish Productivity in Reservoirs

The morphoedaphic index (MEI) has been widely used to estimate fish production and yield in both lakes and reservoirs (Ryder 1982, Jenkins 1982). The index is the ratio of the average total dissolved solids in the lake water and the mean depth of the water body. Generally, what is found is that as MEI increases, fish production and yield increase. That is, fish production is high when TDS is high and/or mean depth is shallow. While the index has good predictive power, the underlying ecological mechanisms remain unclear. It is often assumed that the combination of TDS and mean depth gives an indication of both the abundance of essential plant nutrients through TDS and the ability of the water body to process these nutrients by available light and rapid mineral cycling through mean depth. Thus, it is often assumed that MEI is a good indicator of fish production by predicting pelagic plant production. However, while Adams et al. (1983) found that the MEI did in fact give a reasonable prediction of fish production in 16 midwestern reservoirs, there was no relationship between the MEI and phytoplankton production in these same 16 reservoirs. Thus, while the MEI remains a valuable empirical indicator of fish production, it is unclear how this is mediated. Nonetheless, we can conclude that with a low MEI, reservoir fish production is likely to be low. Youngs and Heimbuch (1982) point out that the most important component of the MEI is mean depth, we can further simplify and state that with increasing mean depth of reservoirs, fish production can be expected to decline (Rawson 1952).

Few facets of reservoir fish ecology have received more comment than the almost universal observation that production of all sorts of fish species is quite high immediately after a water body is impounded. Typically, however, this initial high biomass and production of fish declines within 5 to 20 years and remains at a lower level from that point on (see Ploskey 1981 for a review). Because the decline of production involves sport fish as well as forage and other fish, there is generally a concomitant decline in the success of sport fishing, which the public laments and fisheries biologists attempt to forestall.

There are no natural analogies to the growth of fish in a newly impounded reservoir. Limnologists rarely have the opportunity to study a completely newly formed lake. Most lake formation processes occur during calamitous geologic epochs, such as during or just following glaciation, volcanism, or an earthquake. Even if a limnologist were present at one of these times, in the natural situation it would take fish a relatively long time to colonize a completely new lake, whereas fish are almost always immediately stocked in a newly impounded reservoir. Thus, the period of time of initial high production in reservoirs would probably be long past in a new natural lake before fish naturally colonized it. Thus, the "trophic upsurge" or boom-bust cycle so commonly observed in new reservoirs is unique and has no natural comparison, making the effort to sort through the various proposed explanations for the phenomenon even more difficult.

Because the land that is commonly flooded to make reservoirs often contains soil rich in nutrients, it has been suggested that phytoplankton primary productivity is higher in the beginning of the life of a reservoir than later, when phytoplankton must rely on recycling and inflow from a stream or river. While there is considerable evidence that nutrient concentrations are higher right after impoundment (Dussart et al. 1972, Keith 1975, Nelson et al. 1978), many reservoirs continue to have fairly high nutrient levels long after fish production has declined (McCammon and von Geldern 1979).

Probably of much greater importance to initial fish production in reservoirs is the early productivity of benthic invertebrates, which utilize inundated terrestrial vegetation; many fish, especially small stages of sport fish, feed on these organisms. Aggus (1971) found large concentrations of benthic invertebrates in herbaceous and woody vegetation left in Beaver Lake in Arkansas, while Applegate and Mullan (1967) found few benthnic invertebrates in Bull Shoals Lake, from which standing timber was removed.

It may be of equal importance that newly flooded terrestrial vegetation provides excellent spawning sites for fish. In many reservoirs fish spawn much more successfully and prolifically when structure of this type is available.

Whatever the process, as terrestrial vegetation declines in nutritional value and structural integrity, so does fish production decline. Therefore, the initial high levels of fish production in new reservoirs appear to depend on terrestrial vegetation that provides habitat and food for benthic invertebrates upon which fish feed and structure upon which the fish spawn. With the decline of terrestrial vegetation comes a drop in biomass and production of fish within the reservoir. Fish production then must rely solely on the open-water primary productivity, and for spawning fish must utilize the more resistant structure of the reservoir, such as rocks and cobble.

REFERENCES

Adams, S. M., B. L. Kimmel, and G. R. Ploskey. 1983. Sources of organic matter for reservoir fish production: A trophic-dynamics analysis. *Can. I. Fish. Aquat. Sci.* 40:1480–1495.

Aggus, L. R. 1971. Summer benthos in newly flooded areas of Beaver Reservoir during the second and third years of filling, 1965–1966. Pages 139–152 in G. E. Hall, ed. Reservoir fisheries and limnology. Am. Fish. Soc. Spec. Publ. No. 8.

Aggus, L. R. and G. V. Elliott. 1975. Effects of cover and food on year-class strength of largemouth bass. Pages 317–322 in H. Clepper, ed. Black bass biology and management. Sport Fishing Institute, Washington, DC.

Allan. R. C. amd J. Romero. 1975. Underwater observations of largemouth bass spawning and survival in Lake Mead. Pages 104–112 in H. Clepper, ed. Black bass biology and management. Sport Fishing Institute, Washington, DC.

Applegate, R. L. and J. W. Mullan. 1967. Food of young largemouth bass (*Micropterus salmoides*), in a new and an old reservoir. *Trans. Am. Fish. Soc.* 96:74–77.

Benson, N. G. 1976. Water management and fish production in Missouri River main stem reservoirs. Pages 141–147 in J. F. Osborn and C. H. Allman, eds. Instream flow needs. Vol. 2. Am. Fish. Soc., Washington, DC.

Clady, M. D. and R. C. Summerfelt. 1979. Effectiveness of floating tire breakwaters for increasing density of young largemouth bass in coves of an Oklahoma reservoir. Pages 38–39 in D. L. Johnson and R. A. Stein, eds. Response of fish to habitat structure in standing water. North Central Div., Am. Fish. Soc. Spec. Publ. 6.

Colle, D. E. and J. V. Shireman. 1980. Coefficients of condition for largemouth bass, bluegill, and redear sunfish in *Hydrilla*-infested lakes. *Trans. Am. Fish. Soc.* 100:521–531.

Cooper, W. E. and L. B. Crowder. 1979. Patterns of predation in simple and complex environments. Pages 257–267 in H. Clepper, ed. Predator-prey systems in fisheries management. Sport Fishing Inst., Washington, DC.

Drenner, R. W. and S. R. McComas. 1980. The roles of zooplankton escape ability and fish size selectivity in the selective feeding and impact of planktivorous fish. Pages 587–593 in W. C. Kerfoot, ed. Evolution and ecology of zooplankton communities. University Press of New England, Hanover, NH.

Drenner, R. W., W. J. O'Brien, and J. R. Mummert. 1982. Filter-feeding rates of Gizzard Shad. *Trans. Am. Fish. Soc.* 111:210–215.

Dussart, B. H., K. F. Lagler, P. A. Larkin, T. Scudder, K. Szesztay, and G. F. White. 1972. Man-made lakes as modified ecosystems. SCOPE Rep. 2, Int. Council Sci. Unions, Paris, France. 76 pp.

Fabre, J. H. C. 1913. The life of the fly; with which are interspersed some chapters of autobiography. Dodd, Mead and Company, New York, NY. 477 pp. Tr. by Alexander Teixeiva de Mattos. Library of Congress QL 533 F2.

Fabre-Domerque, P. and E. Bietrix. 1897. Development de la sole (*Solea vulgaris*). *Bull. Mus. Nat. Hist., Paris* 3:57–58.

Ginnelly, G. C. 1971. Investigation of factors limiting population growth of crappie. Fish. Res. Ariz., 1970–1971. 15 pp.

Griffith, J. S. 1978. Effects of low temperature on the survival and behavior of Threadfin Shad, *Dorosoma petenense. Trans. Am. Fish. Soc.* 107:63–70.

Guillory, V. and R. D. Gasamay. 1978. Zoogeography of the Grass Carp in the United States. *Trans. Am. Fish. Soc.* 107:105–118

Hansen, D. F. 1951. Biology of the white crappie in Illinois. *Illinois Natural History Bulletin* 25:211–265.

Hansen, M. J. and D. H. Wahl. 1981. Selection of small *Daphnia pulex* by yellow perch fry in Oneida Lake, New York. *Trans. Am. Fish. Soc.* 110:64–71.

Hassler, T. J. 1970. Environmental influences on early development and year-class strength of northern pike in lakes Oahe and Sharpe, South Dakota. *Trans. Am. Fish. Soc.* 99:369–375.

Heman, M. L. 1965. Manipulation of fish populations through reservoir drawdown, with emphasis on *Micropterus salmoides* (Lacepede). M. A. Thesis, Univ. of Missouri, Columbia, MO. 65 pp.

Howick, G. and W. J. O'Brien. 1983. Piscivorous feeding behavior of largemouth bass: An experimental analysis. *Trans. Am. Fish. Soc.* 112:508–516.

Hunter, J. R. 1979. The feeding behavior and ecology of marine fish larvae. In J. E. Bardach, ed. The physiological and behavioral manipulation of food fish as production and management tools.

Ivlev, V. S. 1961. Experimental ecology of the feeding of fishes. (Translated from the Russian.) Yale University Press, New Haven, CT.

Janssen, J. 1976. Feeding modes and prey size selection in the alewife (*Alosa pseudoharengus*). *J. Fish. Res. Bd. Canada* 33:1972–1975.

Janssen, J. 1978. Feeding-behavior repertoire of the alewife, *Alosa pseudoharengus* and the ciscoes *Coregonus hoyi* and *C. artedii. J. Fish. Res. Bd. Canada* 35:249–253.

Jenkins, R. M. 1982. The morphoedaphic index and reservoir fish production. *Trans. Am. Fish. Soc.* 111:133–140.

Johnson, R. P. 1963. Studies of the life history and ecology of the bigmouth buffalo, *Ictiobus cyrinellus* (Valenciennes). *J. Fish. Res. Bd. of Canada* 20: 1397–1429.

Keith, W. E. 1975. Management by water level manipulation. Pages 489–497 in H. Clepper, ed. Black bass biology and management. Sport Fishing Inst., Washington, DC.

Kramer, R. H. and L. L. Smith, Jr. 1962. Formation of year-classes in largemouth bass. *Trans. Am. Fish. Soc.* 91:29–41.

Lawrence, J. M. 1957. Estimated sizes of various forage fishes largemouth bass can swallow. *Proc. S. E. Assoc. Game Fish Comm.* 11:220–225.

Lerner, E. E., D. J. Hull, D. R. Laughlin, D. L. Wagner, L A. Wilsmann, and F. C. Funk. 1977. Habitat partitioning in a freshwater fish community. *J. Fish Res. Bd. Canada* 34:360–370.

Lillelund, K. 1967. Experimentelle untersuchungen uber den einflub carnivorer Cyclopiden auf die sterblichkeit der fischbrut. *Zeit f. Fischerei (N. F.)* 15:29–43.

Martin, R. G. and R. S. Campbell. 1953. The small fishes of Black River and Clearwater Lake, Missouri. *Univ. Missouri Stud.* 26:45–66.

Martin, D. B., L. J. Mengel, J. F. Novotony, and C. H. Walburg. 1981. Spring and summer water levels in a Missouri River reservoir: Effects on age-O fish and zooplankton. *Trans. Am. Fish. Soc.* 110:370–381.

McCammon, G. W. and C. von Geldern, Jr. 1979. Predator-prey systems in large reservoirs. Pages 431–442 in H. Clepper, ed. Predator-prey systems in fishery management. Sport Fishing Institute, Washington, DC.

Nelson. R. W., G. C. Horak, and J. E. Olson. 1978. Western reservoir and stream habitat improvement handbook. Contract No. 14-16-0008-2151, U.S. Fish Wildl. Serv., Washington, DC. 250 pp.

Ney, J. J. 1981. Evolution of forage-fish management in lakes and reservoirs. *Trans. Am. Fish. Soc.* 110:725–728.

Nikolsky, G. V. 1963. The ecology of fishes. Academic Press, New York, NY.

Noble, R. L. 1981. Management of forage fishes in impoundments of the southern United States. *Trans. Am. Fish. Soc.* 110:738–750.

Ploskey, G. R. 1981. "Factors affecting fish production and fishing quality in new reservoirs, with guidance on timber clearing, basin preparation, and filling." Technical Report E-81-11, prepared by Fish and Wildlife Service, National Reservoir Research Program, U.S. Department of the Interior, for the U.S. Army Engineer Waterways Experiment Station, CE, Vicksburg, MS.

Ploskey, G. R. and R. M. Jenkins. 1980. Inventory of U.S. Reservoirs. U.S. Fish Wildl. Serv., National Reservoir Research Program, Fayetteville, AK. 33 pp. (mimeo).

Rawson, D. S. 1952. Mean depth and fish production of large lakes. *Ecology* 33:513–521.

Ryder, R. A. 1982. The morphoedaphic index: Use, abuse, and fundamental concepts. *Trans. Am. Fish. Soc.* 111:154–164.

Savino, J. F. and R. A. Stein. 1982. Predator-prey interaction between largemouth bass and bluegills as influenced by simulated, submersed vegetation. *Trans. Am. Fish. Soc.* 111:255–266.

Shirley, K. E. and A. K. Andrews. 1977. Growth, reproduction and mortality of largemouth bass during the first year of life in Lake Carl Blackwell, Oklahoma. *Trans. Am. Fish. Soc.* 106:590–595.

Vinyard, G. L. and W. J. O'Brien. 1976. Effects of light and turbidity on the reactive distance of bluegill sunfish (*Lepomis macrochirus*). *J. Fish. Res. Bd. Canada* 33:2845–2849.

Vogele, L. E. 1975. Reproduction of spotted bass in Bull Shoals Reservoir, Arkansas. Pages 1–21 in U.S. Fish Wildl. Serv. Tech. Pap. 84.

Walburg, C. H. 1977. Lake Francis Case, a Missouri River reservoir: Changes in the fish population in 1954–75, and suggestions for management. United States Fish and Wildlife Service Technical Paper 95.

Werner, E. E., D. J. Hall, D. R. Laughlin, D. J. Wagner, L. A. Wilsmann, and F. C. Funk. 1977. Habitat partitioning in a freshwater fish community. *J. Fish. Res. Board Can.* 34:360–370.

Youngs, W. D. and D. G. Heimbuch. 1982. Another consideration of the morphoedaphic index. *Trans. Am. Fish. Soc.* 111:151–153.

Zaret, T. M. 1981. Predation and freshwater communities. Yale University Press, New Haven, CT. 187 pp.

CHAPTER 9

Reservoir Ecosystems: Conclusions and Speculations

ROBERT G. WETZEL

Understanding of operational and functional continuities among biological components is among the foremost objectives of contemporary ecological research. An underlying premise of this compilation of reservoir analyses is that reservoir ecosystems differ in significant ways from natural lake ecosystems. In such analyses I feel that it is of the utmost importance that we not fall into the limnological quagmire of the 1920s–30s. As lakes and their biota were being analyzed and described during this fledgling period of limnology, insightful attempts by a few perceptive workers to understand commonality were long thwarted by hundreds of published reports of often minor structural differences among lakes. The excesses of the lake typologists expanded our knowledge of variations among lakes but simultaneously impeded progress in perceiving the threads of functional unity that exist in aquatic ecosystems.

All of the authors of this review of reservoir ecosystems recognize many functional similarities between man-made and natural lakes. In order to effectively manage and utilize reservoirs, however, it is important to understand the structural differences between these man-made ecosystems and natural lakes while simultaneously appreciating their functional similarities. Understanding of these structural differences is mandatory for effective management and use of impounded water resources. Failure to recognize basic similarities in metabolic functioning and community interrelationships of the biota, however, will only result in redundancies in reservoir research that can be basically understood from our existing knowledge of processes in natural lakes.

In the ensuing remarks I attempt to summarize the major structural differences between reservoirs and natural lakes as delineated, in part, in previous chapters. In addition I emphasize functional biotic similarities; usually the differences are only a matter of extent of process intensities and rates at which they occur. Although the latter statement is also true for distinctions and differences between oligotrophic and eutrophic natural lake ecosystems, one must be aware of the importance of system management and its effects on the resultant processes and their rate functions in reservoirs. Because of the more variable nature of reservoir structure and processes, modified limnological analyses are often required to differentiate effectively the differences and similarities between reservoirs and natural lakes.

CONTRASTS AND SIMILARITIES BETWEEN RESERVOIRS AND NATURAL LAKES

The detailed summary of major characteristics of reservoir ecosystems in the foregoing chapters emphasizes their numerous structural differences from natural lake ecosystems. Other works (e.g., Margalef 1975, Ryder 1978) have approached the problem from a similar direction. I have summarized some of these characteristics and properties in Table 9.1. Many of these contrasts are congruous with the preceding chapter discussions. A few contrasts, however, are in direct opposition to those of others, in particular to Ryder (1978). A latter work was based almost entirely from a fish biology stand-point rather than from one of the ecosystem as an integrated whole. Moreover, a number of Ryder's contrasts, although providing a fertile basis for discussion, are in question on theoretical grounds or have little supportive foundation with data from a spectrum of reservoir ecosystems.

Reservoirs are created predominantly in regions where large natural lakes are sparse or unsuitable (e.g., too saline) for human exploitation. In these regions the climate tends to be warmer than is the case among many natural lakes and results in somewhat higher average water temperature, longer growing seasons, and precipitation inputs that are closely balanced to, or less than, evaporative losses. The drainage basins of reservoirs are consistently much larger in relation to the lake surface areas than is the case among many natural lakes. Because reservoirs are almost always formed in river valleys and in the base of the drainage basins, mor-

Table 9.1 Comparative characteristics and properties between reservoir and natural lake ecosystems

Properties	Reservoirs	Natural lakes
Geographical distirbution	Predominantly southern (N. hemisphere) in non-glaciated regions	Predominantly northern in glaciated regions
Climate	Precipitation often low and evaporation high or greater than precipitation	Precipitation commonly exceeds evaporative losses
Drainage basins	Usually narrow, elongated lake basin in base or drainage basin; area large in comparison to lake area (ca. 100:1 to 300:1)	Circular, lake basin usually central; area usually small in comparison to lake area (ca. 10:1).
Shoreline development	Great, astatic	Relatively low; stable
Water level fluctuations	Large, irregular	Small, stable
Thermal stratification	Variable, irregular; often too shallow to stratify in riverine and transitional zones; often can temporarily stratify in lacustrine zones	Natural regime; often dimictic or monomictic
Inflow	Most runoff to lake via river tributaries (high stream orders); penetration into stratified strata complex (over-, inter-, underflows); often flow is directed along old riverbed valley	Runoff to lake via small tributaries (low stream orders) and diffuse sources; penetration into stratified waters small and dispersive
Outflow (withdrawal)	Highly irregular with water use; withdrawal from surface layers or from hypolimnion	Relatively stable; surface water
Flushing rates	Short, variable (days to several weeks); increase	Long, relatively constant (one to many years)

Table 9.1 ***(continued)***

Properties	Reservoirs	Natural lakes
	with surface withdrawal, disruption of stratification with hypolimnetic withdrawal	
Sediment loading	High with large drainage basin area; flood plains large; deltas large, channelized, gradation rapid	Low to very low; deltas small, broad, gradation slow
Deposition of sediments	High in riverine zone, decreasing exponentially downreservoir; greatest in old riverbed valley; highly variable rates seasonally	Low, limited dispersal; relatively constant rates seasonally
Suspended sediment in water	High, variable; high percentage clay and silt particles; turbidity high	Low to very low; turbidity low
Allochthonous particulate organic matter (POM)	Moderate, especially fine POM during spates and inundation of floodplains	Low to very low
Water temperatures	Somewhat higher (generally more southern climate)	Generally lower (concentrated in more northern climatic regions)
Dissolved oxgyen	Somewhat lower solubilities (higher temperatures); greater horizontal variability with inflow, withdrawal, and POM loading patterns; metalimnetic oxygen minima more common than maxima	Somewhat higher solubilities (lower temperatures); small horizontal variability; metalimnetic oxygen maxima more common than minima
Light extinction	Horizontal gradients (in kilometers) predominate; light extinction irregular and often very high, particularly in riverine	Vertical light gradients (in meters) predominate; variable but relatively low extinction from dissolved organic

Table 9.1 ***(continued)***

Properties	Reservoirs	Natural lakes
	and transitional zones from abiogenic particulate matter; euphotic zone commonly increases in lacustrine zones	compounds and biogenic particulate matter
External nutrient loadings	Generally higher than a natural lakes (larger drainage basin, more human activity, greater water level fluctuations); variable, often unpredictable	Variable but relatively predictable; loadings often moderated by biogeochemical influences of wetland/littoral interface zones
Nutrient dynamics	Horizontal gradients predominate; dependent upon sedimentation rates, residence times, and flow regimes; concentrations in water decrease with distance from headwaters; irregular internal loading	Vertical gradients dominate; often low internal loading, particularly in lakes without severe culturally induced eutrophication
Dissolved organic matter (DOM)	Allochthonous and benthic sources predominate; irregular, often high; refractory DOM predominates	Allochthonous and littoral/wetland sources predominate; relatively constant, often high; refractory DOM predominates
Littoral zone/wetland	Irregular and limited by severe water level fluctuations	Dominates primary production in most lakes; important to regulation of nutrient and dissolved and particulate organic matter loadings
Phytoplankton	Marked horizontal gradients; volumetric primary productivity (or P_{max}) decreases from	Vertical and seasonal gradients predominate; small horizontal gradients; light and

Table 9.1 ***(continued)***

Properties	Reservoirs	Natural lakes
	headwaters to dam; areal primary productivity relatively constant horizontally; light and inorganic nutrient limitations predominate	inorganic nutrient limitations predominate
Bacterial heterotrophy	Pelagic, particle-associated, and benthic bacterial heterotrophy predominates in riverine zones	Benthic and littoral/wetland bacterial heterotrophy predominates in most lakes
Zooplankton	Maximal development common in transition zone; horizontal patchiness high; particulate detritus (including adsorbed DOM) variably augments phytoplankton as food source	Vertical and seasonal gradients predominate; horizontal patchiness moderate; phytoplankton is a predominate food source
Benthic fauna	Low diversity with minimal and irregular littoral zone; productivity low to moderate; initially high with inundated terrestrial vegetation	Moderate to high diversity; productivity moderate to high
Fish	Predominantly warm-water species composition; differences often related to initial stocking; spawning success variable (low with low water levels), egg mortality increases with siltation, larval success reduced with less refugia; productivity initially (5–20 yr) high,	Warm- and cold-water species composition; spawning success good, egg mortality lower, larval success good; productivity moderate

Table 9.1 *(continued)*

Properties	Reservoirs	Natural lakes
	then decreasing. Occasional two-story fishery (warm- and cold-water species) successful, particularly in mountainous reservoirs	
Biotic community relationships	Diversity low; niche specialization broad; growth selection (r) rapid; immigration-extinction processes rapid; net production high soon after inundation, then decreasing	Diversity high; niche specialization moderately narrow; growth selection (K) variable, relatively homeostatic; immigration-extinction processes slow; production low to moderate, relatively constant
Ecosystem succession rate	Similar to lakes but greatly accelerated; greatly stressed by human manipulations of basin and drainage basin	Similar to reservoirs but greatly protracted

phometry of reservoir basins is usually dendritic, narrow, and elongated. These physical characteristics affect biological processes in many complex ways, the most important of which are light and nutrient availability. Reservoirs receive runoff water mainly via high-order streams, which results in high energy for erosion, large sediment-load carrying capacities, and extensive penetration of dissolved and particulate loads into the recipient lake water. Since the inflows are primarily channelized and often not intercepted by energy-dispersive and biologically active wetlands and littoral interface regions, runoff inputs are larger, more directly coupled to precipitation events, and extend much farther into the lake per se than is the case in most natural lakes. All of these properties result in high, but irregularly pulsed, nutrient and sediment loading to the recipient reservoir.

Extreme and irregular water level fluctuations commonly occur in reservoirs as a result of flood inflow characteristics, land-use practices not conducive to water retention, channelization of primary influents, flood control, and large, irregular water withdrawals for hydropower generation operations. Multiplicative effects on loadings result. Large areas of sediments are alternately inundated and exposed; these manipulations usually prevent the establishment of productive, stabilizing wetland and littoral flora. Erosion and resuspension of floodplain sediments augment high loadings from drainage basin sources. Sediments are alternately shifted between aerobic and anaerobic conditions, which enhances nutrient release. The reduction or elimination of wetland and littoral communities around many reservoirs minimizes their extensive nutrient and physical sieving capacities that function effectively in most natural lake ecosystems (Wetzel 1979, 1983).

Within the lake itself the irregular dynamics of inflow and rapid, variable flushing rates markedly alter environmental conditions for biotic communities. A reservoir can be viewed as a very dynamic lake in which a significant portion of its volume possesses characteristics of and functions biologically as a river. Often the riverine portion of a reservoir operates analogously to large, turbid rivers in which turbulence, sediment instability, high turbidity, reduced light availability, and other characteristics preclude extensive photosynthesis despite high nutrient availability. Although phytoplanktonic primary productivity of riverine sections of reservoirs can be high per unit water volume, the limited photic zone reduces areal productivity, as in large rivers (e.g., Wetzel 1975, Minshall 1978, Bott 1983); this reduction is only partially ameliorated by turbulent, intermittent recirculation of algae into the photic zone. As turbidity is reduced and the depth of the photic zone increases in the progression through the transitional to the lacustrine reservoir regions, areal primary productivity increases with greater light penetration and depth of the trophogenic zone. Nutrient limitations, so characteristic of natural lakes of low to moderate productivity, can then occur to varying degrees as losses of nutrients exceed renewal rates. Light limitation is unquestionably a dominant control of productivity in many reservoirs, as it is in many productive natural lakes and rivers. In many cases light limitations in reservoirs result primarily from clay and silt inorganic turbidity. Where light limitations are associated with high dissolved organic matter loadings, as in many tropical and subtropical reservoirs, the quantitative and selective light restrictions are quite analogous in reservoirs and natural lakes.

Internal loading of nutrients, normally low in natural lakes, can be high in reservoirs. Much of the internal nutrient loading is associated with the irregular inflow and withdrawal dynamics, which can disrupt thermal stratification and oxygenation patterns that suppress sedimentary nutrient release and redistribution in physically more stable natural lakes.

Loading of dissolved organic matter (DOM) from allochthonous sources is generally quantitatively similar in reservoirs and lakes. The great sedimentation rates of inorganic particulate sediments in reservoirs could remove significant amounts of DOM from the water, although sorption sites on many of these particles are likely saturated before they reach the lake. Loading of particulate organic matter (POM) from allochthonous sources is normally a minor part of organic carbon budgets in natural lakes (e.g., summary of Wetzel 1983). Fine POM of allochthonous sources can be a significant portion of organic loading in reservoirs, particularly during spates and when floodplains are inundated at high water levels. How significant these two detrital sources (fine POM and DOM absorbed to suspensoids) are to microconsumers faced with limited living particle availability is unclear. Preliminary evidence suggests that the detritus augments but does not supplant photosynthetic sources.

At the present time evidence indicates that bacterial heterotrophic productivity is similar to or only slightly greater in reservoirs than in natural lakes of comparable productivity (Wetzel 1983). As in most natural lakes much of the bacterial metabolism in reservoirs is benthic in sediments rather than pelagic. However, because of the limited wetland and littoral development and irregularly high loading of the fine POM, pelagic bacterial heterotrophy can assume greater significance in reservoirs than occurs in many lakes. How significant particle-associated bacteria are as a supplemental food source for microconsumers of reservoirs is unclear; in natural lakes this source is usually minor. How much pelagic bacterial carbon is directly used by zooplankton in bacterivory vs. direct utilization by protozoan bacteriovores in the "microbial loop" is unclear for reservoirs as well as natural waters. The proportion is probably quite variable on daily as well as seasonal time scales. Moreover, the utilization of algal picoplankton and cyanobacteria by zooplankton, protozoa, and flagellates is also variable among different ecosystems and over time (Stockner and Antia 1986). Much remains to be evaluated before any generalized statements can be made about conditions where the microbial loop assumes importance. It is not a question of whether the microbial pathway is important but rather of its magnitude under different conditions.

Additionally, at present there appears to be no good evidence that algal heterotrophy of DOM is any more significant in light-limited reservoirs than in natural lakes. Algal heterotrophy is energetically inefficient, and these plants are poorly adapted to compete enzymatically with bacteria for these substrates (Wetzel 1983).

Fish biology and productivity are highly variable in dynamic, constantly changing reservoir ecosystems. The high fish productivity commonly observed soon after reservoir formation can be related to higher benthic fauna productivity associated with greater habitat variability and refugia among inundated terrestrial vegetation. However, it has also been related to initially high nutrient and organic matter loadings during the "trophic surge" period. Many reservoirs are not completely cleared for forest and shrub vegetation prior to inundation, particularly in the riverine areas. Although few quantitative data exist on the importance of dead, standing vegetative substrata for "epiphytes" and benthic fauna, many qualitative estimates indicate these substrata and associated fauna can be major food sources for fish (K. W. Thornton, personal communication). As these habitats decay and decline, fish must shift to alternative predominantly pelagic food sources. The high turbidity can decrease visual predation on pelagic zooplanktonic food sources. Fluctuating water levels, high siltation, and heavy predation often result in high egg and larval fish mortality in littoral areas.

The environmental conditions of reservoir ecosystems tend toward large, rapid, and erratic fluctuations. Often insufficient time exists for complete population growth and reproductive expansion to occur before a succeeding major disturbance occurs. These instabilities result in biota that tend to be few and well adapted with broad physiological tolerances (low diversity, less specialization, rapid growth). As in all restrictive, stressed environments, the productivity of the adapted organisms can be high, as high or greater than in more homeostatic lakes.

FURTHER REFLECTIONS

The physical and resulting biotic structural characteristics of reservoirs are indeed markedly divergent from general properties of natural lake ecosystems. Nonetheless, upon careful scrutiny I reach the overriding conclusion that reservoirs are very similar to lakes in terms of basic ecological processes and their controlling factors. Process-regulating parameters function in reservoirs in irregular and extreme ways that greatly confound typical continuity and successional processes of the biota. These extremes restrict the types of organisms in reservoirs to those with broad physiological tolerance ranges and wide behavioral adaptations. However, the basic processes of individual, community, and ecosystem metabolism are the same in both reservoirs and natural lakes. As we learn more about the physiological, behavioral, and regulating properties of both ecosystem groups I am confident that our views of reservoir and lake processes will converge, not diverge.

Predictability is based in biological unity and is a primary objective of limnological understanding. The irregular and extreme variations in physical factors in many reservoirs frustrate our search both for unity and order and for generalized management techniques among these ecosystems. These irregular characteristics place additional analytical burdens upon persons with responsibilities for effective management of reservoir resources. Consequently, much information of individual reservoir properties is required to obtain reasonably sensitive predictability of biotic and water quality responses to irregular inputs and manipulations. Predictive models of biotic responses, developed mainly from a larger database from natural lakes, must be used with great caution when applied to reservoirs. The process responses are the same, but the input variables are more complex and dynamic than in many natural lakes.

Time is a critical factor in nearly all response functions. In order for most of the processes (e.g., stratification, sedimentation, population growth, competitive exclusion, etc.), as we conventionally know them largely from natural lake ecosystems, to come into play and be effective, adequate time is required under relatively undisturbed conditions. As natural or manipulated conditions become more irregular and variable, responses are increasingly incomplete before they are altered or destroyed. The result is an increasingly chaotic succession of responses, reductions in interdependability, and less biotic stability. Although certain physical factors (e.g., precipitation rates) are beyond the control of reservoir management, many factors (e.g., outflow discharge rates, stratification-withdrawal relationships, water renewal rates) can be effectively regulated. An appreciation of the complexities of biotic processes can only lead to much more effective management of multipurpose reservoir ecosystems.

ACKNOWLEDGMENTS

The instructive dialogue with K. W. Thornton on views presented here is particularly appreciated. The support of the U.S. Department of Energy (EY-76-S-02-1599, COO-1599-222) is gratefully acknowledged. Contribution No. 495, W. K. Kellogg Biological Station of Michigan State University.

REFERENCES

Bott, T. L. 1983. Primary productivity in streams. Pages 29–53 in G. W. Minshall and J. R. Barnes, eds. Stream ecology: The testing of general ecological theory in stream ecosystems. Plenum, New York, NY.

Margalef, R. 1975. Typology of reservoirs. *Verh. Internat. Verein. Limnol.* 19: 1841–1848.

Minshall, G. W. 1978. Autotrophy in stream ecosystems. *BioScience* 28:767–771.

Ryder, R. A. 1978. Ecological heterogeneity between north-temperate reservoirs and glacial lake systems due to differing succession rates and cultural uses. *Verh. Internat. Verein. Limnol.* 20:1568–1574.

Stockner, J. G. and N. J. Antia. 1986. Algal picoplankton from marine and freshwater ecosystems: A multidisciplinary perspective. *Can J. Fish. Aquat. Sci.* 43:2472–2503.

Wetzel, R. G. 1975. Primary production. Pages 230–247 in B.A. Whitton, ed. River ecology. Blackwell Scientific Publs., Oxford.

Wetzel, R. G. 1979. The role of the littoral zone and detritius in lake metabolism. *Arch. Hydrobiol. Beih. Ergebn. Limnol.* 13:145–161.

Wetzel, R. G. 1983. Limnology. 2nd Edition. Saunders College Publishing, Philadelphia, PA. 860 pp.

INDEX